U0095289

食 物 小 传

咖啡

Coffee

A Global History

〔英国〕乔纳森·莫里斯 著

赵芳 译

北京联合出版公司
Beijing United Publishing Co.,Ltd.

我的父母安妮·莫里斯和格雷厄姆·莫里斯教我的，关于咖啡的不多，关于人生的却很多。谨以此书献给挚爱的他们。

目录

咖啡是一种风靡全球的饮品。不仅在四大洲[1]有商业种植，还在七大洲有热情消费：连在南极洲的科学家们都爱喝咖啡，国际空间站上甚至也有一台意式浓缩咖啡机。咖啡历经的旅程可谓丰富多彩：从埃塞俄比亚的森林到拉丁美洲的种植园，从奥斯曼帝国时期的咖啡馆到"第三波"咖啡浪潮，从咖啡壶到胶囊咖啡机。

本书是由专业历史学家撰写的首部咖啡全球史。书中解释了咖啡如何获得人们的青睐，而世界各地的咖啡口味差异又为何如此之大。从15世纪阿拉伯半岛第一批喝咖啡的苏非派[2]信徒到21世纪亚洲的精品咖啡消费者，本书不但讨论了咖啡的受众，还讨论了人们喝咖啡的原因与地点、制作咖啡的方式，以及咖啡的风味。书里不仅明确了咖啡种植的区域和方式、农场工人和农场主，以及咖啡豆加工、交易与运输方式；还分析了咖啡背后——涉及中间商、烘焙商和咖啡机制造商——的贸易，并剖析了联系生产商与消费者的结构背后的地缘政治。

---

① 指非洲、亚洲、欧洲和美洲。——译注
② 伊斯兰教的神秘主义派别。——译注

全球咖啡产量的区域分布（单位：%）

年份	非洲和阿拉伯半岛	加勒比海地区	亚洲	拉丁美洲
1700	100	0	0	0
1830	2	38	28	32
1900—1904	1	4	4	91
1970—1974	30	3	6	61
2011—2015	9	1	32	58

本书开篇主要介绍了商品咖啡与精品咖啡的区别，以及这些区别如何决定咖啡从种子变为一杯饮品的过程。

咖啡的历史可分为五个时期。最初，咖啡作为"伊斯

蒸汽朋克咖啡机。这台咖啡机造型独特，由各种升级改造的部件组成，成为2017年伦敦咖啡节上讨论的焦点。其核心是一个冰滴咖啡壶，用它制作一升咖啡要花八个小时。

兰教的酒"供人享用，咖啡树种植在也门的山间梯田，咖啡豆在印度洋和红海沿岸地区的穆斯林之间进行交易。18世纪，欧洲人将咖啡变成一种殖民货物，迫使农奴和奴隶在遥远的爪哇岛和牙买加等地种植咖啡。

19世纪下半叶，巴西的咖啡产量剧增，促进了美国大众消费市场的发展，咖啡转变成一种工业产品。20世纪50年代以后，非洲和亚洲通过种植罗布斯塔咖啡，重新在世界贸易中占据较大份额，咖啡因此成为一种全球商品。罗布斯塔咖啡抗病力更强，但口味也较涩，可以与较便宜的咖啡拼配或做成速溶咖啡。20世纪末，为了反对商品化，人们掀起了一场将咖啡重塑为"精品特饮"的运动，而咖啡跨国界的成功或许将助其迎来史上的第五个时期。

咖啡贸易涉及多种定义及计量单位。历史数据集往往不太有可比性，因此，本书并未强行捏造虚假统一的信息，

咖啡树的形态及解剖图

3

而是原原本本地呈现数据。用于宏观比较的统计数据有多种来源，旨在说明咖啡贸易变化的方向和规模，并且我们不能完全以数字为准。

来杯咖啡吧，再翻开此页，开始享受你的咖啡全球史之旅！

从种子到杯子

咖啡属于日常饮品——人们吃早餐时第一件事便是大饮一口咖啡，上午十点左右的休憩时光也可见它的身影，午后喝咖啡使人提神，晚餐后喝咖啡则有助于消化。对于喝咖啡的人而言，大多能凭直觉判断出自己心目中的好咖啡，不过鲜有人了解这杯好咖啡诞生的成因。本章将探索咖啡从种子到杯子的旅程，同时揭示这趟旅程中让咖啡豆成为商品咖啡抑或精品咖啡的不同选择。

商品咖啡还是精品咖啡？

消费者之所以缺乏品鉴咖啡的知识，主要是因为整个产业将其变成了一种完全同质化的商品，从而掩盖了它的复杂性和多样性。某个时期成批摘果的咖啡豆会和其他时期采摘的咖啡豆混在一起；农场产出的特色各异的咖啡豆会相互组合；来自不同地区的咖啡豆被打包成袋，然后会贴上相同的标签出口海外；生咖啡豆无须人们一睹其貌便可以买卖交易，烘焙后与来自不同国家的咖啡豆拼配，贴上"饱满""芳醇""大师之选"等供人辨识的通用标签进行销售。

此类策略可让不同产地的咖啡相互替代。因为干旱、霜冻、病害等自然事件和战争等人为因素都可能对某一地区的咖啡产量造成长达数年的负面影响。种植者、出口商、中间商和烘焙商都把推动咖啡同质化当作风险管理的策略。对于全球咖啡而言，商品咖啡至少占总产量的90%。

剩下的5%至10%则属于精品咖啡：品质高，风味独特，产地特征明显。和葡萄酒一样，咖啡的风味反映了其种植的品种、产地的微观环境（风土条件）、生长期的主要气候条件，以及摘果、加工、储存和运输时的小心程度。葡萄酒中含有约 300 种影响其风味的分子；而对咖啡来说，估计远不止上千种。过去 30 年间，这种精品［英文为 special(i)ty，欧洲人习惯保留中间的字母"i"，美国人则习惯省略］咖啡的产量呈指数级增长。

咖啡农场可分为三种类型。第一类是大型农商，在巴西拥有大规模的种植园，数量占比不到 1%，但产量约占世界供应量的 10%。第二类是家族企业，主要分布在中美洲和哥伦比亚，数量占 5%，而产量占 30%，其中大都是精品咖啡。第三类是面积小于 5 公顷（约合 12 英亩）的小农场，数量占 95%，产量则占全球总量的 60%。对大多数农场而言，咖啡属于经济作物，都按自给自足的机制种植。[1]

阿拉比卡

咖啡是来自非洲的礼物，那里现已发现 130 多种咖啡属的树种。阿拉比卡咖啡树是从埃塞俄比亚西南部高地及肯尼亚和南苏丹的接壤地区进化而来的，至今仍属于野生树种。目前，阿拉比卡咖啡树的商业种植区全都分布在热带地区。该树种无法在热带以外的地方存活，因为气温一

---

[1] David Browning and Shirin Moayyad, 'Social Sustainability', in *The Craft and Science of Coffee*, ed. B. Folmer (Amsterdam, 2017), p. 109.

旦低于 0℃，咖啡树就会死亡。阿拉比卡是首个——直到20世纪都是唯一一个——为满足人类消费而种植的咖啡品种，如今其产量大约占世界咖啡产量的三分之二。

阿拉比卡咖啡树是一种野外多年生常绿灌木，生长高度在森林半阴树冠下可达9至12米（约合30至40英尺），常被误认成乔木。它可以自花授粉，长出一簇簇散发芳香的白色小花。花的数量和大小主要取决于天气。在阵雨浇淋下会开花，但干旱的生长条件让果实结得最好。在半干旱的气候下，它只有一个花期；而在降雨量较大的地方，可能至少有两个花期，还会有开花结果同时发生的情况。花蕊中的籽结出的果实被称为咖啡樱桃。这些果实在开花后30至35周成熟，等到颜色由绿变为深红（有些品种会变为黄色），便是时候摘果了。

阿拉比卡咖啡树（上图产地为印度），沿侧枝一路开花结果。咖啡树同时开花结果意味着它一年可摘果两次。

咖啡樱桃成熟的各个阶段。果实应在深红色时采摘，过熟须丢弃，如最右边的那种。

埃塞俄比亚森林咖啡。这些半野生的阿拉比卡咖啡树生长在森林树冠下，由小农户种植。这些咖啡豆被誉为"原生种咖啡"，是几个世纪以来不同亚种在野外杂交培育的结果。

每颗咖啡樱桃包含两粒扁平的种子，通常不太严谨地被称为咖啡豆。每粒种子都覆有一层类似于羊皮纸的内果皮，皮下有甘甜、柔软的粉红色果肉保护。偶尔也会出现圆形种子，即所谓的"圆豆"。生产者专门将它们挑出来，高价出售，他们宣称这种圆豆的口感更好，形状也有利于烘焙均匀。鉴于一颗圆豆轻于两颗普通的平豆，有人怀疑圆豆的价格更高是为了弥合利润差距。

罗布斯塔

在 19 世纪的最后 30 年里，一场突如其来的具有毁灭性的叶锈病改变了整个咖啡行业，几乎摧毁了整个亚洲的咖啡生产。咖啡种植者开始在各地寻找替代品种，特别是在荷属东印度群岛[①]。他们尝试过种植大果咖啡，亦称利比里亚咖啡，但最终证明该品种的咖啡树同样易得叶锈病。于是，他们转向了中果咖啡，也就是罗布斯塔，该树种源自刚果，经由比利时移植而来。

罗布斯塔咖啡树不仅抗叶锈病，而且比阿拉比卡更耐温耐湿，因此能够在海拔较低的地区生长。该树种呈伞状，结出的咖啡樱桃较小，但数量较多，成簇分布，更易摘果。罗布斯塔易于栽培，使其得以用于咖啡生产，最近还被引进越南种植。目前，罗布斯塔约占世界咖啡产量的 35% 至 40%。

---

① 今印度尼西亚。——译注

罗布斯塔的一大缺陷便是：产出的咖啡品质不如阿拉比卡。罗布斯塔咖啡的口感通常被形容为烧焦的橡胶味。它几乎总被用来生产拼配咖啡，并且常用于制作速溶咖啡。市面上销售的精品罗布斯塔咖啡（尤其是产自印度的）通常得益于更优良的培育和加工方式。此外，罗布斯塔的咖啡因含量是阿拉比卡的两倍。

罗布斯塔咖啡樱桃（上图产地为印度尼西亚），比阿拉比卡咖啡樱桃更小、更圆，呈大型集中的簇状。

品　种

纵观咖啡历史，大部分时期人们都只种植阿拉比卡的两个品种。一种是铁皮卡，最接近埃塞俄比亚野生咖啡的商业品种。另一种是波旁，在法国人首次种植咖啡的殖民地上发生自然变异的品种。波旁产量较高，与铁皮卡的花香味不同，波旁偏果味。

巴西的咖啡研究人员开发出了矮种卡杜拉与卡杜艾，由于两者产量高且易于培育，在"二战"后大受欢迎。卡杜拉与自然生长的阿拉比卡与罗布斯塔杂交而成的"混种帝汶"再次杂交，培育出另一个产量高的矮种卡蒂姆，卡蒂姆抗病性更强，可以在低海拔地区种植。这些品种深受商品咖啡种植者的喜爱，但咖啡爱好者们对其口感持怀疑态度。

然而，到了21世纪，精品咖啡界对咖啡品种的兴趣大增。这主要受埃塞俄比亚的天然品种瑰夏①影响。它于20世纪50年代被引入中美洲，但由于产量低，鲜有人追捧。然而，2004年，巴拿马博克特的翡翠庄园新主人彼得森家族发现自家咖啡农场的一杯杯咖啡之所以独具风味，全都归功于该品种的咖啡豆。他们将之挑选出来，参加美国精品咖啡协会举办的烘焙师公会咖啡大赛，连续三年它都拔得头筹。2007年，一批精品咖啡的价格是商品咖啡的一百多倍。

瑰夏咖啡豆制作的咖啡具有复杂的芬芳花香，还混合着茶的醇厚口感，如今许多农场都在种植瑰夏，而帕卡马拉、黄波旁等其他品种也广受欢迎。2018年，一种经自然处理的巴拿马瑰夏咖啡以803美元/磅②的价格被拍卖，创造了新的世界纪录，当时商品咖啡的价格仅为1.11美元/磅。③

---

① 英文写作 Geisha 或 Gesha，因同日本的艺伎英文名（geisha），故也有译成"艺伎咖啡"的。——译注

② 磅，英美制质量或重量单位，1磅约为0.45千克。——译注

③ Nick Brown, 'Natural Geisha Breaks Best of Panama Auction Record at $803 per pound', www.dailycoffeenews.com, 20 July 2018.

风土条件

　　培育咖啡的微观环境，或说风土条件，会对其风味和口感产生重要影响。这涉及多种变量，难以孤立区分，若将任何单一因素与杯中咖啡的风味直接挂钩，都可谓不妥。

咖啡树在高海拔地区营养丰富的土壤中生长得最好，通常是火山土壤，如在萨尔瓦多（见上图）。

　　对咖啡树生长影响最大的因素是温度与海拔高度。在气温超过32℃的条件下，阿拉比卡的产量极为不佳，要避免这种情况，海拔高度至关重要。在纬度为10度以内的赤道地区，合适种植的区域海拔通常在900米（约合2950英尺）以上；在亚热带靠热带这端，种植咖啡树的区域的海拔要低得多。最好的埃塞俄比亚咖啡树，如耶加雪菲咖啡树，生长在海拔约1800米（约合5905英尺）处，而夏威夷著名的科纳咖啡带则从海拔 200 米（约合 656 英尺）处起往上延伸。

在咖啡树种植地区，海拔越高，咖啡豆的品质就越好。生长在高处的咖啡豆风味更加浓郁，这大概是因为昼夜温差较大。低海拔地区种植的咖啡豆更柔软，密度更小，老化得更快。萨尔瓦多根据海拔高度对其咖啡进行分级：种植于海拔 1200 米（约合 3937 英尺）以上的才算上品"极高山豆"。

咖啡豆的口感明显受到平均气温的影响，温度较低意味着口感更佳，如酸度（对味觉的刺激）和果味更强，而高温则会导致香气减少及异味增加。针对阿拉比卡主要产区的调研表明，全年气温相对稳定的地区具备种植精品咖啡的理想条件：气温变化范围保持在 13°C 至 25°C。在阿拉比卡种植区，只有四分之一的土地满足这些条件。[1]

咖啡树生长的土壤类型多种多样，前提是需满足土层深厚、排水良好、养分丰富的条件。火山土壤格外受欢迎，有人认为这种土壤培育出来的咖啡豆酸度更高。同样，咖啡树生长的气候条件也多种多样，前提是每年要有差不多 125 厘米（约合 49 英寸）的降雨量。这个雨量既可均匀或季节性分布，也可以通过灌溉系统进行人工灌溉。

咖啡树自然生长于部分阴凉的环境中，只需四分之一的光照量便能使其高效生长。阴凉条件对口感的影响目前尚有争议。哥斯达黎加的研究表明，阴凉的生长环境可以提升咖

---

[1] Charles Lambot et al., 'Cultivating Coffee Quality', in *Craft and Science*, ed. Folmer, pp. 21–2.

啡的酸度，减少苦味和涩味，但哥伦比亚的研究结果恰好相反；而夏威夷的研究发现，同种咖啡树无论是在阴凉环境还是在阳光充足的条件下种植，其豆子的杯测品质并无差异。[①]

培　育

农户从种子开始培育咖啡作物，在树苗生长到约18个月时，将它从盆栽移植到地里。咖啡树苗通常在第3至4年开始结果，第5至7年左右达到商业成熟期。树木长出垂直于树干的侧芽，再从侧芽抽出新枝，逐渐向外生长。果实沿着前一年的侧枝成簇状分布。为方便采摘，农户通

巴西某个大咖啡庄园里的咖啡苗圃

① Shawn Steiman, 'Why Does Coffee Taste That Way', in Coffee: A Comprehensive Guide, ed. R. Thurston, J. Morris and S. Steinman (Lanham, MD, 2013), P. 298.

过修剪，将咖啡树的高度限制在 8 至 10 米（约合 26 至 33 英尺）。

理论上，只要保持健康，一棵树的生产寿命可以无限长。然而，当咖啡树压力过大时（由于缺乏养料或水），为了保护这一年的收成，它真的会自我牺牲，让自己的叶子变黄，树枝枯萎到无法恢复的程度。

阴凉的生长环境可以让咖啡树减轻压力，调节空气和土壤的极端温度，减少树对养料的需求。在没有天然荫蔽的条件下，也可以种植树木来挡风固坡，防止土壤侵蚀。农民经常借助自给作物来为咖啡树遮阴。

希望高产的种植者通常采取密植计划，让咖啡树相互遮阴，形成"咖啡树篱"。这虽使单株产量下降，但却大大增加了每公顷的总产量。该种植方法适用于机械化耕作，常见于巴西大型商业咖啡种植园中。这种"阳光生长"或"经过技术化"的咖啡需要更多肥料和频繁除草，并且更易受疾病侵袭，尤其因为捕食昆虫的鸟类在减少。

咖啡叶锈病很常见，其破坏性最大，是由咖啡驼孢锈菌这种真菌所致。感染后，叶片通常出现橙色和黄色斑点，然后脱落。叶落之后，树枝会枯死，最终导致整棵树死亡。2011 年，中美洲突然暴发叶锈病疫情，5 年内殃及 70% 的咖啡农场，导致 170 万咖啡工人失业。[1]咖啡果小蠹是一种黑色的小甲虫，在咖啡樱桃内部产卵，是咖啡面临的主要

---

[1] 'Applied R and D for Coffee Leaf Rust', www.worldcoffeeresearch.org, accessed 10 December 2004.

昆虫威胁。一旦幼虫孵化，它们便会边吃边开路，直到从咖啡豆中钻出来。严重的虫害可摧毁约一半的作物。

采 收

虽然咖啡的风味受风土条件和品种影响，但所生产批次的咖啡质量却主要取决于采收和处理方式。

咖啡樱桃要么手工采摘，要么剥枝采收。高品质的咖啡只能出自同一批完全成熟的咖啡豆，因此采摘工人要一颗颗挑着采收，将其他未成熟的咖啡樱桃留在树枝上待其成熟。选择性采摘需要大量人力与成本。为此，小生产者需联合起来，相互帮忙采摘咖啡樱桃，大庄园则招募季节工，激励他们在采摘时把控质量。

夏威夷的选择性手工采摘。这些咖啡树属于阿拉比卡咖啡中的铁皮卡品种。

相比之下，剥枝采收是指一手抓住树枝，另一只手顺着枝干将所有果实剥下来。咖啡樱桃（和其他剥落物）直接落到地上或预先放置好的网中，接着分好类以供加工处理。为商品咖啡市场供货的生产者普遍采用剥枝采收法，利用雨季（开花期）的相对可预测性，估算出四分之三左右的咖啡樱桃成熟的时间。

平地的咖啡种植园会利用采收机器进行机械剥枝，沿着咖啡树篱采收。据统计，巴西种植园五名工人三天用机器采收的量相当于危地马拉山区一千名工人手工采摘一天的量。①

自然处理

在处理过程中，需要去除咖啡樱桃的保护层。首先，采用"干"或"湿"处理法去除果核外的果皮和果肉，然后送去研磨，目的是去除豆子表面残留的内果皮。

自然处理，或称"干"处理的过程是，先在阳光下将咖啡樱桃铺于水泥平地上，然后在果实逐渐干枯和腐烂时定期翻动它们，再用耙子将它们堆成一排排的样子，每排大约 5 厘米（约合 2 英寸）高，空出一定间隔。平铺的咖啡樱桃会定期移至旁边空地上，而之前铺上咖啡樱桃的地方则要重新晒干。要让咖啡豆达到可以脱壳的程度，整个

① Oxfam, *Mugged: Poverty in Your Coffee Cup* (Oxford, 2002), p.20.

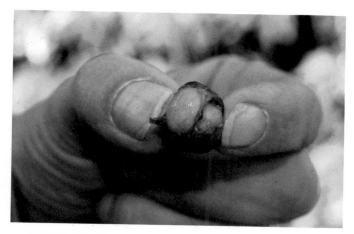

咖啡樱桃的剖面图。每颗种子表面都附着一层黏糊糊的果胶，而种子与果皮之间的粉色柔软果肉起着保护种子的作用。

自然处理的第一阶段：在巴西的露天空地上晒咖啡樱桃。

脱壳前自然干燥的巴西咖啡樱桃

处理过程大约需要两周的时间。

　　自然处理尤其适用于缺水的环境。也门是最早推行此技术的地方，咖啡樱桃通常在种植地所在的山村的屋顶平台上进行干燥。此过程增强了咖啡的醇厚度和果味，使其回味中带有一丝"野生"的味道。处理得好的咖啡豆，制作出来的风味令人振奋；处理得不好的，则会让人想到农场。

　　自然处理的主要吸引力在于其成本效益。大多数商品咖啡都采用干处理法，包括巴西"天然培育"的阿拉比卡，还有世界上几乎所有的罗布斯塔。从咖啡品质的角度而言，干处理面临的困难在于无法保持同等质量。采收的咖啡樱桃通常都只是事先进行人工分拣，这增加了坏咖啡樱桃通过筛选以及污染其余批次的风险。整个处理过程也不稳定，果实经历了各种温度，存在过度发酵或发霉的风险。

水　洗

　　湿处理法制作的咖啡往往更加醇和，味道更稳定，酸度更佳。采用此处理法时，要先将咖啡樱桃置于浮选池中。饱满成熟的咖啡樱桃会沉到池底，而漂浮在池面的则是过熟和未熟的咖啡樱桃，以及树枝和树叶等残余物。这些漂浮物都会被清除，然后用泵送系统将池底的樱桃输送到脱皮机中。所有咖啡樱桃被同一筛网挤压脱皮，果皮与果肉分离，从而将豆子筛除出来。而未熟的咖啡樱桃太过坚硬，无法通过筛网脱皮，也会在此时被清除掉。

夏威夷浮选池面浮着未熟和过熟的咖啡樱桃——湿处理法的准备阶段

埃塞俄比亚所谓的非洲高架网床上晾晒湿处理后的咖啡豆

然后咖啡豆被倒入清水池中，在此放置 12 至 14 个小时发酵，直到豆子表面黏稠的果胶层分解。抓一把豆子，感受一下，通常便能做出评估：等果胶层在水中完全分解，便可以将咖啡豆打捞上来。

湿处理法需要用到大量的水，而为了减少用水，人们近年来开发了脱皮机和脱胶机。将脱胶机直接放到脱皮机筛网后面，然后让筛除出来的豆子相互摩擦，除去彼此的果胶层，避免用水来发酵。

等除去果胶层，再将咖啡豆倒入水中冲洗，然后使其干燥到剩下 11% 的水分含量。这个过程可以在空地或桌上完成，有时是在透明的塑料棚内，目的是利用温室效应，保护咖啡。在气候条件起决定作用的情况下，还会使用干燥机。

半日晒与蜜处理

20世纪80年代，巴西发明了半日晒处理法。照这种处理方法，咖啡豆可经脱皮机筛除出来，但之后连带着果胶直接送到空地上晾晒。由此生产的咖啡豆风味甚佳，既有干处理后的醇厚度，也不乏湿处理后的酸度。中美洲的咖啡生产商采用了此处理法，通过改进而开创了蜜处理法，让咖啡豆在更潮湿的环境中慢慢干燥，使其逐渐散发香气。此方式生产的咖啡豆可分为黄蜜、红蜜和黑蜜，三种的区别在于咖啡豆上保留的果胶量——黑蜜保留了全部的果胶，需要长达30天的时间进行干燥。[1]

动物处理

印度尼西亚咖啡又名猫屎咖啡（或称努瓦克咖啡），出了名的经由棕榈麝香猫处理而成。它们通常会食用落在地上的咖啡樱桃，然后排出无法消化的咖啡豆，实际相当于为其进行了脱皮处理。接着，这些豆子以常规方式被收集、清洗和制成成品。据说，猫消化系统中的酶素让排出的咖啡豆具有一种独特风味，猫屎咖啡的零售价因而高达100美元/磅，而在纽约，一杯猫屎咖啡的售价为30美元。果然，其他国家如今也发现了具备同样功能的动物，包括越南黄

---

[1] 黄蜜保留了60%的果胶，红蜜保留了75%的果胶。——译注

鼠狼、泰国大象和巴西雅库鸟。

　　令人难过的是，对猫屎咖啡的热烈追捧驱使人们大量捕获和囚禁动物，对其强行喂食咖啡豆。经过野生处理认证的咖啡确实存在，但市面上出售的大都是假借这个名义，有时还会进行化学处理。让动物处理咖啡，最重要的是进去什么便出来什么，生产商可以采用更便宜的咖啡樱桃来降低成本。猫屎咖啡通常酸度低、苦味轻——而这两种品质更多的是指调和口感差的咖啡，而非提升好口感。

巴厘岛某市场的猫屎咖啡摊位前有两只棕榈麝香猫

养　豆

　　经过脱皮处理和干燥，咖啡豆在脱壳前最好避免恶劣天气的影响，在内果皮中"养"上一两个月。这个过程可以让豆子进一步成熟，散去刚处理后残留的青草气息。

在印度西南部的马拉巴尔地区，受季风和湿气影响的露天仓库可用来养豆。在这期间，咖啡豆会逐渐变成金黄色，水分含量增加到13%至14%，随即膨胀起来。这相当于19世纪漫长的海上旅程给咖啡带来的影响，由此诞生了一种低酸度的芳醇咖啡。

脱壳前，干燥过的咖啡豆在内果皮中养豆。

脱 壳

经过养豆，咖啡豆表面残留的内果皮可通过脱壳机除去。然后在倾斜的分拣台上将咖啡豆按目数大小划分等级。咖啡豆的标准目数是以直径1/64英寸为计算单位（通常为

10目至20目）。① 例如，肯尼亚的 AA 级咖啡豆的大小是18目。在脱壳过程中，要同时除去有缺陷的咖啡豆，可手工或用色选机来识别那些未熟、破损或被虫蛀的豆子。

脱完壳后的咖啡生豆可装入黄麻袋中保存。这种麻袋可让豆子免受日晒雨淋，同时方便空气流通，避免豆子发霉。不过，空气流通可能会导致咖啡豆变味，而麻袋里以石油为原料的涂层会使豆子带有所谓的"袋子"味。近年来，精品咖啡生产商开始在咖啡生豆封入麻袋前，先将豆子装进大容量的多层塑料袋中，防止气体和水分入侵。标准麻袋所装的重量是 60 千克（约合 132 磅）。就咖啡豆贸易而言，许多统计数据都是以麻袋为单位，而非按重量计算。

一台正在运行的分拣机。咖啡豆会随着分拣台的倾斜和振动被层层筛选，因此它们是按密度来划分等级的。

---

① 目数是指每平方英寸筛网上的圆孔数目，这些圆孔通常以 1/64 英寸为单位，所谓 10 目的咖啡豆，就是指直径为 10/64 英寸的咖啡豆。——译注

分　类

当一袋袋咖啡豆到达航运港口时，中间已经被多次转手。水洗处理厂与脱壳厂对于资本投资和产量都有一定要求，小农户通常负担不了。在一些产地，几乎所有供应链环节都掌握在私人手中，种植者在农场门口便把咖啡豆卖给了中间商；在其他产地，农民合作经营水洗处理厂，最后统一将咖啡豆卖给脱壳厂。对于处理过的咖啡豆而言，出口代理商是在原产国的最后接收者，他们准备好咖啡豆，将其打包，运往海外消费市场。

在供应链的每个阶段，根据生产国采用的标准化类别，单个批次会被拼配成更大的批量。例如，巴西主要依据样品豆所允许的瑕疵率，将咖啡豆分为七个等级：衡量的瑕疵是其中掺杂的过熟、损坏或遭病害的咖啡豆，以及石头、表皮、细枝等数量。这些宽泛的等级划分可与国际市场上的现行价格挂钩，从而使咖啡豆能够作为一种商品进行交易。

国际贸易

咖啡国际贸易主要通过两大咖啡期货交易市场进行：纽约洲际交易所（ICE）和伦敦国际金融期货和期权交易所（LIFFE），两者交易的咖啡豆品种分别是阿拉比卡和罗布斯塔。这两大期货交易所进行的标准合约，都是提前确定未来某一特定时间交割一定数量的主流咖啡豆。纽约洲际交易所交易的标准合约属于"C型"咖啡期货合约：交割货物数量为 1.7 万千克（约合 3.75 万磅）口味温和的阿拉比卡水洗咖啡豆，在 350 克（约合 12 盎司）样品豆中，瑕疵等级为 9 至 23。瑕疵较少的批次溢价交易，而低于标准的批次则折价出售。"C型"咖啡期货合约包括了 20 个咖啡生产国，其中包括哥伦比亚在内的一些国家可自动溢价，而其他国家，如多米尼加共和国，则会折价。在两年的交易期里，共有 10 个交割月份的报价，在交割月份前一个月关闭交易仓位。

这些期货交易所在咖啡贸易行业扮演着重要角色。由于交易所记录的是期货经纪人对未来咖啡豆交割所约定的所有合约，因此会出现一个价格指标。根据交易所的规则和标准合约，其中一些（尽管很少）合约的交割方式是咖啡豆的实物交割。实物咖啡豆更多还是通过远期合约进行交割，以标准期货价格为基点，在基点的基础上计算某一特定等级咖啡豆的最终交割价格。因此，投标的价格要考虑到买卖差价，比如纽约洲际交易所 10 月以 +10 的报价收进 12 月交割、从 y 国运来的 x 级咖啡豆（即 12 月交割的"C型"咖啡期货合约价格"增加 10 个点"）。

这让最终交易价格具有不确定性，因为交易所的期货和期权市场可以抵消或对冲这个价格。咖啡期权交易（指买卖的选择权，而非债券，在一定的价格起点或有效期内按期货合约进行交割）始于 1986 年的纽约洲际交易所。它使咖啡市场涌进了大量的投机者，以至于在 2015 年，合约交易量是全球阿拉比卡咖啡产量的 27 倍。[1]对咖啡贸易商而言，咖啡生豆实物量与交易量之间的差距使其获利，因为其中带来的资金流动意味着期货经纪人更容易对冲头寸。他们把咖啡价格波动所带来的风险从自身转移到了金融投机者身上。相反地，投机成分的增加也带来了更大的价格波动，使产业价值链中的其他人受影响，如无法进入期货市场的种植者和小型烘焙商。

20 世纪 80 年代，在伦敦国际金融期货和期权交易所交易罗布斯塔咖啡豆的期货合约经纪人们。

---

[1]　Eric Nadelberg et al., 'Trading and Transaction', in *Craft and Science*, ed. Folmer, p. 207.

运　输

咖啡期货经纪人通过进出港现货市场的交易，获得了对咖啡生豆的控制权。商品咖啡市场由全球几大贸易商构成的关系网主导。最大的贸易商是位于汉堡的诺伊曼咖啡集团，处理全球 10% 的咖啡需求。贸易商要负责管理咖啡豆的物流运输，如仓储和航运，并将其交付给烘焙商，后者通常是以期权合约的形式购买咖啡豆。咖啡豆用集装箱运输，每箱大约有 275 袋。发往同一家烘焙商的所有货物可直接装入一个单独的集装箱，咖啡豆通过空气动力被吹进集装箱，等运到目的地时再被吸入筒仓。

美国进口咖啡豆的主要港口分布在纽约和新奥尔良（2005 年，新奥尔良因卡特里娜飓风来袭而闭港，引发了美国将面临咖啡短缺的短暂性恐慌）。欧洲进口咖啡豆的头号港口是安特卫普，欧洲大陆进口的咖啡约一半都储存在那里。

脱咖啡因

脱咖啡因的工厂通常位于港口腹地。咖啡生豆首先要进行蒸汽处理，然后浸泡在热水里，使其膨胀起来。这个过程可使加入的溶剂——二氯甲烷或乙酸乙酯——发挥效力，去除咖啡豆中的咖啡因。然后，排空溶剂，低价出售咖啡因提取物，以蒸汽清洗并干燥处理咖啡豆。在高压条件下，也可采用液体或超临界二氧化碳脱咖啡因。这种方式虽然成本高，但需要去除的其他化合物较少。另一种方式是采用瑞士开发的水处理法，以热水作为溶剂。先将咖啡生豆在热水中浸泡八小时左右，然后滤干咖啡豆，热水冲过含活性炭的滤床滤掉咖啡因。这样便得到了浓缩水，接着将咖啡豆重新浸泡其中，使其重新吸收剩余的风味。这种脱咖啡因处理法通常成批进行，一批咖啡豆过滤出来的水再返给下一批过滤。

含咖啡因和脱咖啡因的咖啡生豆样品

烘 焙

咖啡烘焙的零售业务主要掌握在大型跨国食品集团手

中，如雀巢、JDE、盛美家（旗下咖啡品牌有福仕杰）、卡夫亨氏（旗下咖啡品牌有麦斯威尔）和奇堡。中等规模的烘焙商通常为超市零售商和连锁零售店生产自己品牌的咖啡，并为餐饮业（饭店、餐馆和宴会餐饮）供货。小型烘焙商往往专营精品咖啡，包括单一产地和种植园咖啡，而规模极小的烘焙商则生产小批量咖啡，销售渠道主要是自家商店和当地供应网。

咖啡烘焙的基本原理是让咖啡豆均匀受热，并将温度最终加热到 200°C 至 250°C。大多数烘焙商使用的都是所谓的滚转炉，金属炉内的咖啡豆在火上不停被搅动。滚转炉通常分批次烘焙，每批持续 8 至 20 分钟，烘焙时间取决于对成品的需求。大型烘焙商使用的兴许是可以不间断作业的流化床式烘焙机，以高压热气烘豆，烘焙时间大约两分钟。

在烘焙过程中，随着咖啡生豆渐渐失去水分，淀粉焦糖化并转化为糖分，豆子的颜色也逐渐变黄，然后变成浅棕色。当温度加热到大约 205°C 时，会听到咖啡豆在自身内部气压积聚作用下爆开的声音，即所谓的"第一爆"。咖啡豆表面的棕色继续加深，直到温度加热至 225°C 左右，豆子的细胞壁爆开，即"第二爆"，渗出带有光泽的油脂，覆盖到豆子表面。此时若继续烘焙，咖啡豆剩余的糖分会碳化，豆子的颜色则会变为黑色。

咖啡的风味主要取决于两大因素——咖啡豆的烘焙程度与烘焙速度。烘焙师会通过咖啡豆从烘焙机中发出的声音和气味，以及取样器里烘焙好的豆子的外观，判断如何

滚转炉刚烘焙好的咖啡豆被倒至冷却盘中

调整机器。不少烘焙机可以通过编程完成特定的烘焙指令。

人们认为，浅度烘焙——慢慢烘到"第一爆"与"第二爆"之间出豆——能使大多数精品咖啡呈现最佳风味。若进一步烘焙，香气、风味和酸度都会随之减弱，而醇厚度和苦味则会增强。在开始变苦前，糖分逐渐焦糖化，豆子的甜度在"第一爆"与"第二爆"之间会达到峰值。"第二爆"之后，烘焙过的风味会盖过咖啡豆本身的味道，这就是通常选择深度烘焙罗布斯塔等低品质咖啡豆的原因。

烘焙一旦完成，必须尽快冷却咖啡豆，避免继续烘烤：大多数滚转炉会将烘好的豆子倒入多孔托盘中，并在托盘中搅拌，让冷空气通过，使其散热。另外，也可以朝烘好的咖啡豆喷水或将之浸在水中。由于咖啡豆会吸收水分，商业咖啡烘焙师通常会借冷却的名义，以此增加咖啡豆的重量（因此也增加了价值）。

速 溶

速溶咖啡，又称可溶性咖啡，是通过冷冻干燥或喷雾干燥制作而成的。用175℃的高压热水从烘焙和研磨后的咖啡豆中萃取原液，将原液浓缩后冷冻到零下40℃。冷冻成冰的原液分解成小颗粒，并在加热的真空室中干燥。颗粒中冷冻的水分直接升华为蒸汽，由冷凝器排出。喷雾干燥是年代更早、成本更低的处理方式，需要将浓缩的咖啡提取物喷入干燥塔中，咖啡提取物通过250℃的热气流时则会变成脱水粉末。

拼 配

通过拼配，烘焙师可以为自身品牌开发出独具一格的口味，也可控制成本。咖啡拼配通常精选未水洗的巴西阿拉比卡咖啡豆（通常叫作自然豆或桑托斯咖啡豆），这种豆子的口感相对适中，因此可以成为拼配咖啡的主要组成部分。然后，混合哥伦比亚咖啡豆等更具特色的所谓柔和豆，使拼配咖啡整体更有个性。面向大众市场的低成本拼配咖啡首选罗布斯塔咖啡豆，然后加入桑托斯咖啡豆，并以其他产地的豆子作为补充。高品质的拼配咖啡通常先将每批咖啡豆烘出合适的甜度，再混合不同品种的豆子；然而，烘焙前将咖啡豆拼配好，成本效益更高，因此面向大众市场的烘焙师采用的是一种通用的拼配方法。

烘焙师的技巧在于调整现有咖啡豆的烘焙与拼配，从而使某一品牌的口味保持一致，即便实际采用的咖啡豆已发生变化。这通常需要用一种咖啡豆代替另一种：某些烘焙师与咖啡期货经纪人签订远期合同，而交割的咖啡豆会根据供应情况而变化。例如，一份罗布斯塔咖啡豆合同可以允许采用乌干达标准级或象牙海岸二级的咖啡豆。

杯 测

烘焙后，便可以鉴别咖啡豆经过种植和处理后呈现的风味。烘焙师通过杯测检查产品最终的品质及口感的醇厚度。同样，"杯测"也可用于评估咖啡期货经纪人所提供的样品，原产国的买家和生产商也会使用此法。"比较杯测"是比较某一样品烘焙器中同一大小包装的咖啡生豆，因此每包准备的试样都相似。将研磨好的咖啡豆倒入瓷杯，由杯测师评估干燥的咖啡样品的香气，然后再加入热水，由杯测师检测其湿香气。等待四分钟，杯测师可用咖啡匙"破渣"——撇开表面浮末，鼻子尽量凑近杯子，捕捉咖啡的香气，然后一匙匙地饮咖啡，让舌头上的味蕾尽量感受弥漫开来的咖啡口感。

如今，精品咖啡界已采用国际咖啡品质研究所（CQI）的 Q 级杯测体系，使杯测师的评估规范化。所有得到 Q 级认证的杯测师都要经过培训和考核，确保评估标准的一致性。杯测师既要对咖啡的香味／香气（干／湿）、酸度、风味、醇厚度、余味、统一性、干净度、均衡度、甜度以及整体印象进行评分，还要对瑕疵扣分，最后得出总分（满分 100 分）。总分在 80 分以上的咖啡被视为精品咖啡。

包　装

在离开烘焙厂之前，咖啡豆可以有长达三天的时间脱气（排出烘焙时积累的二氧化碳），然后用密封袋或真空袋包装好。为了方便消费者，大多数咖啡豆都会在工厂里预先研磨好，尽管这样会增加暴露在氧气中的咖啡表面积，导致开封后咖啡加速变质。品级更高的咖啡豆通常密封在装有单向排气阀的咖啡袋里，在避免空气进入的情况下，豆子得以继续脱气。还有一些咖啡采用充氮包装，可完全排出袋里的氧气。

冲　泡

冲泡咖啡的操作及设备数不胜数，大致可归为四种基本冲泡方式：

咖啡冲泡设备。从左到右依次为：爱乐压、法压壶、60度锥形手冲滤杯、手冲壶和虹吸壶。

1. 煮沸——将咖啡粉和水加热煮沸，如土耳其咖啡的制作方式。

2. 浸泡——在咖啡粉中倒入热水浸泡一定时间，如使用法压壶的制作方式。

3. 滤泡——用热水直接冲泡咖啡粉，如冲滤的制作方式。

4. 气压冲泡——热水以高压穿过咖啡粉，加速萃取咖啡，如浓缩咖啡的制作方式。

咖啡行业有句老话，咖啡从田地到杯子里需要一年的时间，而消费者只需一分钟便能将它搞砸。为避免出现这样的情况，请认真阅读本书的食谱篇。

健　康

对于咖啡这种饮品而言，所选择的冲泡方式是决定其最终咖啡因含量的因素之一。水接触咖啡的时间越长，吸收的咖啡因就越多。其他主要的决定因素还包括饮品分量，而最重要的是罗布斯塔咖啡豆在拼配中的比例，因为这种豆子的咖啡因含量是阿拉比卡的两倍。美国政府公布的《2015—2020 年美国居民膳食指南》指出，一杯标准的 8 盎司（约合 235 毫升）滴漏咖啡含有 96 毫克咖啡因，一杯同样分量的速溶咖啡的咖啡因含量为 66 毫克，而 1 盎司（约合 30 毫升）单份意式浓缩咖啡则含有 64 毫克咖啡因。这种平均值之间的差异很大，但值得注意的是，该指南还指出每天摄入咖啡因含量不超过 400 毫克，则可视为在健康范围内。[①]

作为一种兴奋剂，咖啡因可以增强大脑功能，帮助对抗困倦。大脑活动通常受腺苷调节，腺苷与神经细胞表面的受体结合，能够诱发睡眠，并使血管扩张以获取更多氧气。当咖啡因通过血液进入大脑时，会与神经细胞的受体结合，起到阻止腺苷与其接触的作用，同时收缩血管，缓解头痛。此时，身体会分泌更多肾上腺素，头脑更加清醒，而多巴胺再吸收的速度减缓，快感增强。因此，对于喝咖

---

① U.S. Department of Health and Human Services and U.S. Department of Agriculture, *2015–2020 Dietary Guidelines for Americans*, 8th edn (2015), p. 33, www.health.gov/dietaryguidelines/2015/guidelines.

啡的人而言，这种饮品可以在早晨帮助他们"开启"新的一天，使其上夜班时保持头脑清醒，缓解压力造成的头痛，或者仅仅让他们感觉状态良好。

但短时间内摄入过多的咖啡因可能有害健康。它会使心率加快、血压升高，同时出现由咖啡因引发的"神经紧张"，并伴有头晕、焦虑、失眠和腹泻等症状。关键是，究竟多少才算过量？大脑中咖啡因浓度的峰值会在喝完咖啡后一小时左右出现，而咖啡因在体内的半衰期通常是三到四小时。咖啡因的代谢因人而异，体重、遗传、性别和生活方式等因素都对其具有重要影响。过度摄入咖啡因不但会让人产生依赖性（因而"咖啡瘾君子"不喝咖啡便会头痛），同时也会对其作用产生耐受性。

单人份咖啡中的咖啡因含量不同，且个人对它的反应也有差别，这就可以解释为何几乎无法就饮用咖啡的适量水平提出任何有用的建议。大众健康调查受扰于得到的问题反馈。某一个人一杯咖啡的分量、拼配和冲泡方式都不同于另一个人，而这甚至还未考虑加奶和加糖的影响。

目前研究发现，喝咖啡有不少积极影响。咖啡中所含的化学成分在预防肝脏疾病，保护肾脏、肠道，以及在较小程度上抵御乳腺癌等方面具有一定作用，并且由于咖啡含有大量的抗氧化物，可能会降低患阿尔茨海默病和帕金森病的风险。同时，咖啡已被证明并不具有脱水或利尿作用，或许有助于对抗成年型糖尿病。最近的研究表明，喝咖啡的人要更长寿。《美国医学杂志》的编辑指出，咖啡爱好者应"享受这种轻微但或许有益的瘾"。[1]

20世纪初，一个贝都因人在外约旦利用随身携带的平底锅在火上烘焙咖啡豆。旁边被称为达拉的阿拉伯咖啡壶，底部扁圆，因此可以稳稳立在沙子上。

[1] Joseph Alpert, 'Hey Doc, is it ok for me to Drink Coffee?', *American Journal of Medicine*, CXXII/7 (2009), pp. 597–8.

咖啡经济的开端

咖啡的"起始传说"深受商人喜爱——从前有一天，年轻的埃塞俄比亚牧羊人卡尔迪注意到，自家的羊吃了某种灌木的红色浆果后，变得异常激动。于是卡尔迪亲自尝了尝那种浆果，最后竟然"手舞足蹈"起来。后来或许是一位伊玛目无意间发现了卡尔迪的状况，又或许是卡尔迪去问伊玛目怎么回事，总之伊玛目也去尝了那些浆果。最后出现了两种情况：（1）他发现了那些浆果可以使他在深夜祈祷时保持清醒，所以把它们制成了液体，与他人分享；（2）他厌恶地将之扔进火里，不承想闻到了它们散发出的烘焙香气，于是决定再把它们从余烬中找出来，将之磨碎，加入热水，便可以享用最终制成的饮品！

卡尔迪的故事最早于1671年出现在欧洲，见于安东尼奥·福士托·奈罗尼发表的关于咖啡的论著。奈罗尼是来自黎凡特（今黎巴嫩）的马龙派基督徒，此前已移居罗马。这个故事很可能是他在家乡听闻的。人类首次饮用咖啡的具体时间、地点和形式无法确定。有传言称是在古代遗址中发现了烧焦的咖啡豆，还有人认为波斯医学家、哲学家伊本·西拿（980—1037，拉丁名阿维森纳）在《医典》中描述的草药和汤药取自咖啡树。

可以肯定的是，在有记载咖啡存在的大约前两百年里，即1450至1650年，饮用咖啡的几乎全是穆斯林，正是他们喝咖啡的习惯支撑起了以红海为中心的咖啡经济。该饮品的现代形态都由伊斯兰世界发展而来，当代咖啡馆的模式也建立在此基础上。

奥罗莫族①部落占据了埃塞俄比亚南部的长片土地，包括阿拉比卡咖啡的起源地咖法与布诺地区，他们利用咖啡树的不同成分制作了各种咖啡食物与饮品，其中包括库提（由轻度烘焙过的咖啡嫩叶煮成的茶）、霍茄（以咖啡樱桃的干果皮与牛奶混合而成的饮品）和布纳卡拉（将晒干的咖啡豆与黄油和盐烘焙而成的有刺激性味道的固体零食），可在探险中携带并食用以增强能量。

2017年，埃塞俄比亚国家咖啡博物馆展出的画描绘了卡尔迪的故事。展板前，一位女迎宾员等用咖啡招待大家，她右边名为耶贝纳的黑色陶土咖啡壶里正煮着咖啡。

　　此地最有名的咖啡食物是布纳。它是用咖啡豆的皮做成的饮品，需要在沸水中文火煮15分钟。现如今，咖啡种

① 奥罗莫族（Oromo），生活在埃塞俄比亚奥罗米亚（Oromia）地区，以奥罗莫语为母语。——译注

植户已经开始销售一种与之类似的产品，名叫卡斯卡拉——咖啡樱桃加工过程中所除去的干果皮冲泡而制成的果茶。此饮品最初是由咖啡樱桃加工后全部的干燥残余物——果皮、果肉和果核——制作而成的。

15世纪中叶，在阿拉伯语中被称为咖许的咖啡饮品，似乎是跨越了红海南端的曼德海峡（全长32千米，约合20英里）。它被也门的苏非派用于迪克尔仪式，仪式开始时，苏非长老会从大器皿中用长柄勺舀出一种被称为咖瓦的刺激性药水，在教徒中分发，同时嘴里念诵经文。在整个仪式中，咖瓦扮演了非常重要的角色，因为践行苏非派神秘主义（Sufism）的是白天需要工作的在俗教徒：Sufism这个词源自"Suf"，意指减少欲望，可能是说要少睡觉。

咖瓦最初取材于咖弗塔叶，是一种名叫咖特的植物的叶子。咖瓦有致幻之效，能增强兴奋感，而饮用咖许则有助于祈祷者保持清醒。据传，最早提倡喝咖许的正是苏非派穆夫提①穆罕默德·达巴尼，他于1470年去世。达巴尼是我们目前能与咖啡联系起来的第一位历史人物。阿拉伯学者阿布德·卡迪尔·加齐里在1556年左右写的手稿《为何支持合法利用咖啡》中，提供了咖啡在伊斯兰世界传播的主要信息来源，再次提及了达巴尼曾到过埃塞俄比亚的说法。加齐里写道：

① 穆夫提（mufti），负责解释伊斯兰教法的学者。——译注

他发现人们饮用咖瓦，不过他当时对其特性一无所知。回到亚丁后，他感到身体不适，这时想到了（咖瓦），喝了后便有所好转。于是他反应过来，咖瓦的一大作用就是驱除疲倦，为身体带来某种精神与活力。因此，当他成为一名苏非派信徒时，他和亚丁的其他苏非派信徒便开始饮用这种饮品。[1]

另一种说法是，尽管阿里·伊本·奥马尔·沙吉利在穆哈被誉为咖瓦之"父"，但这种饮品由咖特制成。

那时的亚丁······达巴尼时期，没有咖弗塔，因此他对那些追随他的信徒说："咖啡豆······使人保持清醒，来试试由它制成的咖瓦吧。"信徒们试了试，发现具备同样的功效······而且成本低，制作简单。[2]

起初，咖瓦仅仅指宗教药水，但后来特指只用咖啡豆制成的阿拉伯咖啡，而咖许仍指以干果和香料制成的饮品。

苏非派的宗教活动有助于将咖啡知识传播到北边的阿拉伯地区，即红海东岸的汉志[3]，包括吉达、麦加和麦地那。

---

[1]　Ralph Hattox, *Coffee and Coffeehouses: The Origins of a Social Beverage in the Medieval Near East* (Seattle, WA, 1985), p. 14.

[2]　同上，p. 18.

[3]　汉志（Hijaz），位于沙特阿拉伯王国西部红海沿海地带，境内有麦加和麦地那两座伊斯兰圣城，是伊斯兰教和早期伊斯兰文化的发祥地。——译注

咖啡最终在 16 世纪的某个时期到达了马穆鲁克苏丹国统治下的首都开罗，在此，第一批饮用咖啡的是来自爱资哈尔伊斯兰大学的也门学生。除在宗教仪式上被饮用以外，咖啡逐渐成为一种社交饮品，推动了其在近东地区的传播。

1516—1517 年，奥斯曼帝国征服了埃及，促进了咖啡在这个土耳其人帝国的传播，1534 年，咖啡传到了大马士革，1554 年再传到伊斯坦布尔。两位分别来自大马士革和阿勒颇的叙利亚人——哈卡姆和谢姆——在帝国首都开了两家咖啡馆。两家店都位于市中心，靠近港口与中心集市，吸引了一批来自精英阶层的顾客，其中包括在文人朋友面前分享自己最新作品的诗人、玩西洋双陆棋和国际象棋等游戏的商人，以及坐在豪华沙发或地毯上交谈的奥斯曼帝国官员。谢姆的咖啡馆经营得颇为成功，据说，三年后他带着赚到的 5000 金币回到了阿勒颇。

不过，阿拉伯咖啡和土耳其咖啡的颜色明显不同。人们过去（和现在）喝到的阿拉伯咖啡（咖瓦）是一种半透明的浅褐色液体。它的豆子要先经过浅烘焙，再冷却、碾磨，并和姜根、肉桂，特别是小豆蔻等香料混合。接着将这些混合物置于铜底壶，用水煮上大约 15 分钟，然后倒入一个加热过的、通常带有长壶嘴的小容器，即达拉咖啡壶中。通常主人会为每位客人倒上一小杯（咖啡杯称为芬詹）。

相比之下，奥斯曼土耳其人喝的则是一种不透明的深色饮品，它被一位当代诗人形容为"睡眠和爱情的黑色敌人"——这便是今天的土耳其咖啡卡赫的前身。这种咖啡的豆子要烘焙到颜色变黑，然后碾磨成粉。再将研磨好的咖啡粉与水一起放入切兹韦壶（在土耳其以外被称为伊布里克壶或布里基壶）中，这是一种宽底敞口壶，壶两端宽，中间窄。煮沸后，将壶从火上移开，然后用匙把咖啡表面的泡沫舀入杯中。接着可以将咖啡重新煮沸（至少一次，通常是两次），往杯中倒入更多咖啡，同时试图保留表面的泡沫结构。

伊斯坦布尔的伊玛目利用咖啡豆的烘焙过程来论证饮

制作土耳其咖啡的传统器具便是切兹韦壶，一种小敞口壶，配有长手柄，可以直接放在火上煮。这种壶嘴可以直接将咖啡倒入"芬詹"咖啡杯中。为了方便上咖啡，另一种方式是把煮好的咖啡倒进伊布里克壶，例如上图中埃及人用的这种壶，其特点是带有细长的壶嘴和高高的锥形颈。在土耳其以外的地方，ibrik 这个词的不同说法均可用来形容这两种容器，因为大家认为 cezve 这个词太难发音了。

用咖啡是违禁行为，因为咖啡豆在烘焙中的碳化意味着这种饮品是由无生命的（因而被禁止的）物质制成的。1591年，伊斯兰教谢赫（Sheik ul-Islam，伊斯兰教最高权威）博斯坦扎德·穆罕默德·埃芬迪颁布了一项法特瓦[①]，明确宣告这种饮品取自植物原料，在这之前并没有发生完整的碳化。据一位当代编年史学家称：

> 乌里玛、谢赫、维齐尔[②]等大人物，全都在喝这种饮品。它受欢迎的程度甚至让维齐尔建了极好的咖啡馆作为投资，然后开始将其出租，日租金为一或两个金币。[③]

咖啡馆的出现让新的社会互动形式成为可能。从前，招待他人需要邀请对方来自己家，设宴款待，宴席可能由仆人准备，席间男主人还会展示自己的财产（妻子大概也得现身），凡此种种都形成了主客之分。如今，人们可以在咖啡馆与同等地位的人见面，并通过请对方喝杯咖啡的简便之举，在更平等的基础上款待对方。早期那些咖啡馆的布局有利于促成平等的氛围，因为顾客会按照到达咖啡馆的先后顺序，而非阶级高低，来选择坐在长凳还是墙边的长沙发上。

---

① 法特瓦（fatwa），伊斯兰教令。——译注
② 维齐尔（vizier），伊斯兰教国家对官廷大臣的称谓。——译注
③ Bernard Lewis, *Istanbul and the Civilization of the Ottoman Empire* (Norman, ok, 1963), p. 133.

上图是由荷兰插图家扬·路肯于 1698 年创作的版画，整个作品展现了奥斯曼帝国中围绕咖啡进行的性别区分。画上的男人们正在土耳其的一家咖啡馆里吸烟，而女人们则是私下聚在一起喝咖啡。该画据旅行者的描述而作。

这种形式使那些条件较差的人能够和精英阶层一样招待客人，表现自己的慷慨大方。1599年，一位前往开罗的游客注意到：

> 当士兵走……进咖啡馆，因为付一枚金币而必须找零钱时，他们一定会把那一枚金币花光。在他们看来，将零钱放回口袋然后离开的行为很不妥。换言之，那是他们向平民炫耀自己的方式。而他们的大方则是请彼此喝咖啡，用一杯咖啡给朋友们留下大方的印象，而四杯咖啡也才卖一帕拉（便士）。[1]

在伊斯坦布尔，咖啡馆颇受欢迎，据称在1564年，即最早那批咖啡馆开业十年后，此地营业的咖啡馆已有50多家；到了1595年，据说已多达600家。[2]这个数字可能在某种程度上算上了酒馆和售卖博萨的店铺，同时也可能反映了一个现实——这些商家所经营的业务之间的界限逐渐变得模糊。咖啡馆允许消费可能有问题的东西，还可以游戏和赌博。到了1565年，曾欢迎最早一批人来伊斯坦布尔开咖啡馆的苏丹苏莱曼一世，开始下令关闭阿勒颇和大马士革的酒馆、博萨店和咖啡馆。人们在这些场所"不断以娱乐消遣、违反法律及禁令的行为来消磨时间"，这使他们无

---

① Hattox, *Coffee and Coffeehouses*, p. 99.
② Ayse Saracgil, 'Generi voluttari e ragion di stato', *Turcia*, 28 (1996), pp. 166–8.

法"履行其宗教义务"。[1]后来的两任苏丹塞利姆二世（1566—1574年在位）和穆拉德三世（1574—1595年在位）则进一步颁布了更多、更严厉的法令。

这些法令似乎并未起到太大作用，尤其因为执法官员与民兵本身就是这些场所的常客，其中还不乏经营者。咖啡馆的成功反映了奥斯曼帝国在社会与政治结构上的变化。原本中央集权、等级分明的行政管理模式已被一个权力分散、精英分化、宗教与世俗意识形态彼此竞争的社会所取代。在咖啡馆，一个人可以和其他任何人直接交流、公开对话，这里成为这种全新文化的象征。

正是由于这种暗含的进步性，咖啡馆受到了宗教及政治保守派的攻击。苏丹穆拉德四世1623年即位时尚未成年，难以树立起自身权威，因而建立了一个高度反动的政权。他通过线人网络悄悄监视各个咖啡馆，窃听不利于自身的闲言碎语。1633年，伊斯坦布尔的五个区同时毁于一场大火，据说是咖啡馆里有人吸烟引起的，总之穆拉德下令关闭该市所有类似的经营场所。这个命令同样下达到了奥斯曼帝国的其他城市，如埃尤普市——

> 要求管辖者收到命令后，派人毁掉辖区内所有咖啡烘焙窑，今后不允许任何人开设。自此，凡经营咖啡馆者，一律于咖啡馆门口处以绞刑。[2]

---

[1] Ayse Saracgil, 'Generi voluttari e ragion di stato', *Turcia*, 28 (1996), p. 167.
[2] 同上。

16 世纪末至 17 世纪初，烟草传入土耳其，虽然吸烟似乎才是导致穆拉德发怒的主要原因或借口（据说他曾在夜间微服私访，对犯禁者进行即决审判[①]），但正如一位帕夏所指出的，咖啡馆经营者当下面临的困境是：

> 在咖啡馆里，经营者无力强制要求顾客（其中许多是士兵）不吸烟；人人口袋里都自带烟草，可以随时拿出来就抽。由于（吸烟者）拥有国家公职人员的特权，咖啡馆经营者和城市其他居民无力反对他们。[②]

17 世纪 50 年代中期，伊斯坦布尔城内仍然禁止经营咖啡馆，不过城外的还在公开营业，就像当时在帝国更偏远地区可能仍有人在开咖啡馆。到了 17 世纪最后 25 年，咖啡馆再度出现在了伊斯坦布尔，而在前往奥斯曼帝国的旅行者中，不止一位曾提及咖啡馆在地理位置上的中心地位，这些地点包括开罗的街市、穿越阿拉伯半岛的商队路线以及伊斯坦布尔的公园。

咖啡在整个伊斯兰世界的传播建立起了一个复杂的长途贸易网络，汇集于开罗，并向整个奥斯曼帝国辐射，并

---

① 即决审判（summary justice/judgment），指不经开庭审理而直接作出判决。——译注

② Ayse Saracgil, 'Generi voluttari e ragion di stato', *Turcia*, 28 (1996), p. 167.

阿米德奥·普雷齐奥萨，《君士坦丁堡的土耳其咖啡馆》，1854年，铅笔与水彩画。普雷齐奥萨在伊斯坦布尔生活了整整40年。画上左后角有煮咖啡的壁炉，周围是制作咖啡的设备。画面中，一位波斯商人正在招待壁炉附近的同伴喝咖啡，而在右前方的位置可以看到一个芬詹咖啡杯，有人正握着它的金属杯柄。不过，大多数顾客都更喜欢抽烟。

也门马纳哈附近的咖啡梯田。位于高地的也门村庄周围都是梯田，其中一些种植着咖啡树。村庄的平屋顶上可以晾晒咖啡樱桃。

最终扩展到欧洲。起初，来自埃塞俄比亚的野生咖啡经干燥处理后，会从泽拉（今索马里北部，与吉布提交界处）运往欧洲。它会在此与原产于印度和远东的香料一同作为货物被运至红海，而卸货的港口都是已经有人在喝咖啡的地区。关于咖啡货物最早的历史记录是在1497年，当时它还属于商人从西奈半岛南端的图尔运来的部分香料。[①]

直到16世纪40年代，埃塞俄比亚仍然是唯一的咖啡产地，但由于该帝国北部基督徒与南部穆斯林之间的冲突，咖啡出现了需求不断增长而供应却不稳定的状况，于是人们开始在也门内陆位于帖哈麦沿海平原和首都萨那之间的山区种植咖啡。除了自给作物，农民在自家土地上种植埃塞俄比亚野生咖啡的小型变种。这些地方便是世界第一批咖啡农场。在将近两个世纪的时间里，该地区一直是唯一的商业咖啡生产中心。整个山区分布着村落，村民住在灰泥涂刷的房屋里，四周是石壁围筑的梯田种植区，雨季过后肥沃的土壤会从干谷[②]归返至梯田。到了18世纪初，这些山区养活了大约150万人。

连接这些生产者和最终消费者的链条一如既往漫长而分散。交通运输极其困难，山区与低地市场之间只有羊肠小道。种植者会将干燥后的咖啡樱桃带到最近的城镇换取

---

① Michel Tuchscherer, 'Coffee in the Red Sea Area from the Sixteenth to the Nineteenth Century', in *The Global Coffee Economy in Africa, Asia, and Latin America, 1500–1989*, ed. William Gervase Clarence Smith and Steven Topik (Cambridge, 2003), p. 51.

② 干谷（wadi），北非和阿拉伯沙漠地区干涸的沙砾河床，只有到雨季来临时才有水流。——译注

布和盐等货物。然后这些咖啡豆会通过各种中间商，最终被运至沿海平原的贝特·法奇，此地是主要的批发市场。商人在此购买咖啡并将之存放于仓库，再由骆驼商队运到摩卡（Al-Makha，又称 Al-Mocha 或 Al-Mokka，欧洲人称之为 Mocha）和荷台达的港口，然后对外航运。这些商人大多为印度散居移民，来自古吉拉特[①]的港口城市苏拉特，主导着印度洋周围的贸易。与此同时，他们还控制着也门的信贷网，可能是咖啡种植的主要出资方和实际发起者。

尽管咖啡贸易存在分散性，但它却产生了可观的利润，尤其是栽德派的领袖们获利颇丰，因为内陆地区的部落都忠于他们的领导，这就大大增加了奥斯曼帝国统治的阻力。1638 年，在伊玛目卡西米的领导下，奥斯曼帝国被迫撤离也门，栽德派首次统一了这个国家，并控制了泽拉，从而真正垄断了也门和埃塞俄比亚对全球市场的咖啡供应。在后来的贸易中，凡是来自这两个产地的咖啡豆均被称作"摩卡"，因为它们是从同一个港口一起出口的。

栽德派起义的成功导致了红海地区贸易的重组。原本运往奥斯曼帝国消费的咖啡被阿拉伯三角帆船从荷台达运到了吉达。奥斯曼帝国强行将吉达作为转口港，并将其所得收入用于支持圣地。而从苏伊士运来谷物的船只又满载咖啡回到开罗。开罗商人早在 16 世纪 60 年代就开始定期开展咖啡贸易，他们会将咖啡运往奥斯曼帝国在地中海的

---

① 古吉拉特（Gujarat），印度最西部的一个邦，濒临阿拉伯海，自古以商业发达而闻名。——译注

主要中心城市，如萨洛尼卡、伊斯坦布尔和突尼斯。17世纪50年代以后，咖啡便被运往亚历山大，由控制通往西欧港口通道的马赛商人收购。

与此同时，摩卡也是通往其他咖啡消费地区的重要港口——主要是波斯湾、阿拉伯海和印度洋周围的伊斯兰地区。因此，它也成为整个红海地区与印度贸易最重要的转口港。为了在贸易中占有一席之地，英国东印度公司早在1618年便在此建立自己的仓库，将各种被称为"cowa""cowhe""cowha""cohoo""couha""coffa"①的货物转运到波斯和莫卧儿帝国的代理商（经纪人）手中，这比咖啡在英国流通的时间早了30多年。随着欧洲和中南半岛②之间交往日益频繁，欧洲公司曾在17世纪想方设法将大部分香料贸易转移到自己手中，但咖啡依旧主要集中于穆斯林的商业网。

当时，欧洲人面临的部分问题便是咖啡供应持续的不可预测性。也门山区的农业结构令种植者难以针对市场需求作出相应的调整。作家、旅行家、将咖啡引入马赛的商人之子让·德拉罗克记录了1709年和1711年两次从布列塔尼的圣马洛港到摩卡港的贸易之行。他的记录透露出，尽管法国人请了位印度咖啡经纪人替他们收购咖啡豆，导致贝特·法奇的咖啡价格上涨，但把咖啡豆装满船舱却花了

---

① 以上均为咖啡的各种叫法。——译注
② 中南半岛：又称中印半岛、印中半岛、印支半岛，位于中国和南亚次大陆之间，西临孟加拉湾、安达曼海和马六甲海峡，东临太平洋的南海，为东亚与群岛之间的桥梁。——译注

奥尔弗特·达佩尔，《摩卡港口》，1680 年。达佩尔基于商人和传教士的描述创作了这幅版画。在这幅画中，既可以看到正前方荷兰国旗在荷兰东印度公司的船只上飘扬，也可以明显看到左岸荷兰东印度公司的工厂。

六个月。他们曾遇到的一位荷兰代理预计需要整整一年才能收购一趟航运的货物。到了 18 世纪 20 年代，红海地区的咖啡出口量已经达到每年 1.2 万至 1.5 万吨——这实际上是全球咖啡的供应量。[1] 在接下来的 100 年里，这个数字基本没发生什么变化，不过到了 1840 年，这个量就只占世界产量的 3% 了。考虑到这一点，再加上欧洲人越来越接受这

① Michel Tuchscherer, 'Coffee in the Red Sea Area from the Sixteenth to the Nineteenth Century', in *The Global Coffee Economy in Africa, Asia, and Latin America, 1500–1989*, ed. William Gervase Clarence Smith and Steven Topik (Cambridge, 2003), p. 55.

种饮品，他们试图建立多个可供替代的咖啡种植中心也就不足为奇了。

18世纪20年代后，荷兰人转向了爪哇岛，法国人则转向了加勒比海，因此他们在摩卡和亚历山大的采购量都在下降。不过英国人和美国人加大了采购量，填补了这个落差。也门的咖啡贸易收入依旧可观，因此追求扩张的埃及统治者穆罕默德·阿里试图征服此地，控制其咖啡贸易收入。这促使英国人在1839年夺取了亚丁，维护他们在该地区的势力，并在1850年将其建为自由港。由于不收关税，再加上有深水码头和仓储设施，亚丁超越摩卡成为该地区咖啡的主要出口港。如今，摩卡的海港区有一个小渔船队以及许多遗迹，并且要通过淤塞的河道才能驶入港口，据说这是因为19世纪的美国船只在装载咖啡之前卸下了压舱物。

红海地区咖啡经济衰退的主要原因是消费者——绝大多数为穆斯林——的口味发生了变化。最大的影响因素是19世纪初，印度和伊朗都转向了茶叶贸易，这些传统的东方市场都不再是咖啡的天下了。在埃及，茶可能更受欢迎，茶叶就来自本国境内种植的茶树。20世纪上半叶，凯末尔的土耳其现代化计划包括了让土耳其转变为一个饮茶的国度，以当地作物制成的茶饮代替昂贵的进口饮品。直到西方咖啡连锁店到来才刺激了土耳其咖啡馆文化的复兴。

反之，在过去的两个世纪里，埃塞俄比亚才是咖啡经济真正扩张的国家。19世纪下半叶，埃塞俄比亚皇帝曼涅里克用出口咖啡的收入购买枪炮，它们在1896年著名的阿杜瓦战役中派上了用场。埃塞俄比亚击败了意大利，维护了自己在列强瓜分非洲大陆后作为唯一独立的非洲国家的地位。在埃塞俄比亚，西南部的锡达莫、咖法和季马等奥罗莫族王国生产"野生"咖啡（可能是为了满足向埃塞俄比亚帝国进贡的需求而在农民土地上种植的），东部的哈拉尔地区附近还建立了新的种植园，栽培的品种是经世界各地的种植区培育而来的阿拉比卡咖啡变种。这些咖啡的豆子较大，被称为摩卡长果，以区别于也门（和埃塞俄比亚）最早的"摩卡"咖啡豆。

埃塞俄比亚北部的科普特基督徒也开始种植和消费咖啡。20世纪30年代，年轻的海尔·塞拉西一世[①]依靠咖啡收入来加强权威。然而，他却无力阻止意大利法西斯对埃塞俄比亚的入侵，亚的斯亚贝巴和阿斯马拉的意式咖啡吧便属于意大利在此留下的殖民产物。埃塞俄比亚如今仍是为数不多的咖啡种植国之一，本国的咖啡消费量占自身产量的比重相当大（约50%）。

[①] 海尔·塞拉西一世（Haile Selassie，1892—1975），埃塞俄比亚帝国末代皇帝，同时也是一名基督徒。——译注

让-艾蒂安·利奥塔尔,《用早餐的荷兰少女》,约1756年,布面油画。到了18世纪中叶,咖啡加牛奶和糖成为欧洲资产阶级的标准早餐饮品。画上的咖啡壶被称为三脚咖啡滴壶。将磨好的咖啡放入壶中,倒入热水,然后通过底部的壶嘴倒出冲泡好的咖啡。然而,此设计很容易导致咖啡渣将壶嘴堵住,使咖啡从壶中滴出,故得此名。

［第三章］

殖民时代的货物

17 世纪中叶以前，欧洲鲜有人喝过咖啡，除了奥斯曼帝国统治下的那些人。咖啡传入欧洲后，咖啡馆和咖啡店也应运而生，席卷了欧洲社会的大部分地区。18 世纪，为了满足消费者日益增长的需求，荷兰共和国、法国和英国等欧洲国家开始在亚洲及加勒比海地区的殖民地种植咖啡树，致使咖啡生产的中心发生了结构性的巨变。

咖啡文化在欧洲的传播

这并非一个欧洲人被咖啡豆征服的简单故事。巧克力、咖啡和茶一个紧接着一个来到欧洲大陆，导致其消费者的喜好来回变化。各种复杂的行业规章制度给商人建立销售和供应咖啡的场所带来了阻力。因此，咖啡文化在欧洲传播的不连续性显而易见。威尼斯或许是第一个出现冲泡咖啡的欧洲城市，但咖啡馆直到一个世纪后才在那里开业。伦敦虽然是欧洲最早经营咖啡馆的地方之一，但英国人却一直是最不活跃的欧洲咖啡生产者之一。消费偏好从巧克力转到咖啡的法国人反而后来居上，在 18 世纪的咖啡消费与殖民生产中占据了主导地位。

咖啡文化在欧洲的传播[1]

地区	领土内首次出现咖啡的记录	首次商业航运	首家咖啡馆开业	殖民地首次种植咖啡树
意大利	1575年威尼斯	1624年威尼斯	1683年威尼斯	
荷兰	1596年莱顿 1616年阿姆斯特丹	1640年阿姆斯特丹	1665年阿姆斯特丹 1670年海牙	1696年爪哇岛 1712年苏里南
英国	1637年牛津	1657年伦敦	1650年（？）牛津 1652年伦敦	1730年牙买加
法国	1644年马赛	1660年马赛	1670年马赛 1671年巴黎	1715年留尼汪 1723年马提尼克
德国		1669年不来梅	1673年不来梅 1721年柏林	
哈布斯堡帝国	1665年维也纳		1685年维也纳	

　　信仰基督教为主的欧洲对咖啡的接受程度，反映了欧洲大陆与信仰伊斯兰教为主的近东地区之间的复杂关系。正是欧洲爆发的对"东方"的迷恋，才激发了人们对咖啡的兴趣，然而 17 世纪早期的旅行者在写作中往往试图重构咖啡的过去，从而切断它与穆斯林的联系。意大利人彼得罗·德拉瓦勒[2]认为咖啡乃忘忧药的主要成分，即荷马

---

① 数据源自文献研究。
② 彼得罗·德拉瓦勒（Pietro della Valle，1586—1652），意大利作曲家、作家，曾游历过亚洲。——译注

史诗中海伦用来酿酒的药物。英国人亨利·布朗特爵士[①]称它是斯巴达人在战斗前饮用的黑汤。他们将咖啡与古希腊人联系起来，实际上是通过这种方式宣称咖啡属于欧洲文明，同时提醒同时代的人注意奥斯曼帝国境内那些喝咖啡的基督徒。不过，关于教皇克莱门特八世曾在17世纪喝过咖啡并将其作为基督教饮品施以洗礼的传闻纯属无稽之谈，尽管该故事广为流传的程度说明了那些与咖啡贸易有利益关系的人倒是希望他曾这样做。

1575年，威尼斯已有了咖啡。当时该城一名土耳其商人被谋杀，其库存清单上记录有咖啡制作设备，此即确证。到了1624年，运往威尼斯的咖啡由药剂师作为医药产品出售。1645年，似乎已有店铺获许经销咖啡豆。[②]之后，意大利的其他城邦也开始买卖咖啡。1665年，托斯卡纳获得咖啡的垄断贸易权。或许正是保护药剂师贸易的规章制度，导致直到1683年威尼斯才出现第一家获准供应咖啡的店铺。到了1759年，威尼斯当局被迫将咖啡店数量控制在204家以内——而不到四年，这个上限便被突破了。

可能是类似的限制模糊了咖啡在哈布斯堡帝国早期的历史，特别是在与奥斯曼人相邻以及经常受其侵犯的地区。1665年，土耳其派代表团前往维也纳签订和平条约，随行的人中有两名专门负责准备咖啡的人员。而到了1666年代

---

① 亨利·布朗特爵士（Sir Henry Blount，1605—1682），英国作家、旅行家。——译注
② Markman Ellis, *The Coffee House: A Cultural History* (London, 2004), p. 82.

表团离开维也纳时，据说当地的咖啡贸易正蓬勃发展。维也纳的咖啡贸易掌握在了亚美尼亚人手中，如约翰内斯·迪奥达托，他在 1685 年获得了维也纳首个制作与销售咖啡——经营咖啡馆——的许可证。[1]

另一个与之相悖且被渲染过的故事是，在 1683 年奥斯曼帝国围攻维也纳期间，格奥尔格·弗朗茨·科尔斯希茨基作为奥斯曼帝国背后的间谍，从撤退的土耳其人那里获得了一袋袋被他们遗弃的咖啡豆作为奖励，于是他利用这些咖啡豆将咖啡引进到了维也纳。身着土耳其服饰的科尔斯希茨基刚开始是在维也纳周围兜售提前煮好的咖啡，同时请求当局允许他开设自己的店铺。获准后，他便以"蓝瓶"为招牌创立了那家著名的咖啡馆。1697 年，科尔斯希茨基去世三年后，获得许可的行会"咖啡师兄弟会"成立。[2]该行会的主要创新之处便是在咖啡中加入牛奶，让顾客用咖啡颜色表来选择想要的咖啡浓度。这就是卡布奇诺咖啡的起源，其颜色与嘉布遣会[3]修士腰间系的白色腰带的颜色一模一样。除了增加甜味（或掩盖咖啡的味道），牛奶还具备象征意义——将代表穆斯林的黑色咖啡转化为基督徒的白色甜品。

---

[1]　Bennet Alan Weinberg and Bonnie K. Bealer, *The World of Caffeine* (New York, 2002), pp. 74–9.

[2]　Karl Teply, *Die Einführung des Kaffees in Wien* (Vienna, 1980); Andreas Weigl, 'Vom Kaffeehaus zum Beisl', in *Die Revolution am Esstisch*, ed. Hans Jürgen Teuteberg (Stuttgart, 2004), p. 180.

[3]　嘉布遣会（Capuchin），天主教方济各会的一支，因其会服带有尖顶风帽（capuche）而得名。——译注

此图大约创作于1900年，描绘了科尔斯希茨基身着土耳其服饰在蓝瓶咖啡馆端咖啡的情景，进一步渲染了那些关于他将咖啡引入维也纳的传闻。

英国是欧洲最早形成咖啡馆文化的国家。将咖啡豆带到英国的主要还是来自奥斯曼帝国的移民。1637年5月，牛津大学贝利奥尔学院的希腊学生纳撒尼尔·科诺皮奥是英国有史以来第一个喝到咖啡的人。有时有人认为来自黎凡特的犹太男仆雅各布1650年在牛津开了家咖啡馆，但即便这个人真的存在，那他更有可能是在为主人的朋友们上咖啡，而非卖咖啡。但毫无疑问的是，来自奥斯曼帝国士麦那城（今伊兹密尔）的亚美尼亚人帕斯夸·罗塞在1652至1654年的某个时间点开设了伦敦乃至欧洲第一家有史料记载的咖啡馆。这门新生意的发展可谓相当迅速；1663年，光是伦敦市官方登记在册的咖啡馆经营商便有82位。

罗塞的生意始于伦敦金融城中心圣迈克尔教堂庭院里的一个摊位。金融城位于大伦敦市中心，属于独立的行政区，聚集了伦敦绝大多数的金融与商业机构。商人们会从附近的皇家交易所来到罗塞摊位前，在篷子下边喝咖啡边继续交流。根据1654年关于罗塞生意的最早记载，它提供的是"一种土耳其式的饮品，用水混合某种浆果或土耳其豆制作而成，微烫，味道有些难闻，（但）回味甘醇，有时会导致频繁放屁"[1]。

曾有传单盛赞"咖啡饮品之美。帕斯夸·罗塞首次面向大众在英国制作并销售咖啡"。该传单指出，咖啡是"一种简单纯洁之物"，其制作方法是

磨成粉，加入泉水直至煮沸，煮到大约半品脱[2]的量便可饮用……在能下嘴的前提下，尽量煮热些……如果有机会观察，则会发现咖啡可驱赶睡意，适合办公时饮用，因此晚饭后应避免喝咖啡，除非你打算保持清醒，因为它会在三四个小时内妨碍睡眠。据观察，在咖啡作为大众饮品的土耳其，人们不受结石、痛风、浮肿或坏血病的困扰，肌肤清透净白。咖啡既非泻药，亦非止血药。[3]

---

[1] Ellis, *The Coffee House*, p. 33.

[2] 品脱，容量单位，1品脱约为600毫升。——译注

[3] *The Vertue of the Coffee Drink* (London, undated, possibly 1656), now in the British Library.

说到好处，这种饮品主要是让身体"保持警醒"。比起"低浓度啤酒"或淡艾尔酒——在水质通常比较恶劣的城市里的传统提神饮品，咖啡的好处不难判断。于是咖啡馆很快便取代了酒馆，成为人们谈生意的主要公共场所。

当地的酒馆老板对罗塞的成功大为不满，抱怨他抢走了他们的生意；可是由于罗塞交易的并非酒类，因此他们无法借口称罗塞侵犯了他们的特许权，而是以罗塞非英国公民为由质疑其贸易权，最终促使罗塞与杂货商公司成员克里斯托弗·鲍曼合伙。两人携手将咖啡生意转移到了一些可俯瞰教堂庭院的房间里。尽管没有任何记录表明罗塞在1658年以后还参与经营，但招牌上依旧可见罗塞的侧影，即众所周知的"土耳其人头像"。鲍曼于1662年去世，之后他的遗孀接管了这家咖啡店，直到店铺在1666年的伦敦大火中被烧毁。

在英国内战结束后的克伦威尔统治时代，咖啡馆的出现并非偶然。当时，各个行会的权力被削弱，社会盛行平等主义与严肃清醒的文化价值观念，恰好适合推行一个无酒精的场所用于社交，让每一位顾客得到平等的招待。早期经营者在咖啡馆里布置的是长桌，让人人都有座位，毫无等级之分。咖啡先在火上煮好，然后倒进咖啡壶里，再由服务员为顾客倒入杯碗（即杯具）中。

1660年君主制恢复后，咖啡馆幸免于难，因为反对议会制度的保皇党也曾利用此地之便，在不受监控的情况下交谈。1666年，克拉伦登伯爵向枢密院提议关闭咖啡馆，

托马斯·罗兰森，《咖啡馆里的疯犬》，1809年。这幅讽刺版画虽然是在咖啡馆的繁荣期过了很久后创作的，但却描绘了伦敦咖啡馆的诸多特征。画中有位体态丰满的女主人（在场唯一的女性）在圆形吧台后主持大局。左边壁架上堆放的咖啡壶看上去像是土耳其的切兹韦壶。右边墙上贴着航运与股票经纪人的通知。那条疯犬可能在为顾客进行娱乐表演。

威廉·考文垂则提醒他，"在克伦威尔统治时代……比起在其他地方冒险交流，国王的朋友们更愿意选择利用咖啡馆的言论自由"[1]。牛津也同样如此，1656年，药剂师亚瑟·蒂利亚德在此开设了英国有记载的在首都之外的第一家咖啡馆。蒂利亚德"之所以选择在牛津开咖啡馆，是因为受到了一些当下正在此生活的保皇党人以及自诩大师或才子的人的鼓励"[2]。

① Ellis, *The Coffee House*, p. 73.
② Brian Cowan, *The Social Life of Coffee: The Emergence of the British Coffeehouse* (New Haven, CT, and London, 2005), p. 90.

"大师"一词形容的是那些充满求知欲的绅士，他们渴望了解文化方面的新奇、稀有之物，以及与弗朗西斯·培根等人有关的经验主义、类似于科学探究的新兴领域。大师们并非朝臣，可以自由地了解各种新现象，并在这些所谓的"便士大学"（咖啡馆一杯咖啡的价格以"便士"计）里讨论。

蒂利亚德的顾客里有"近代物理学之父"艾萨克·牛顿、天文学家埃德蒙·哈雷（"哈雷彗星"便得名于此人）和收藏家汉斯·斯隆（斯隆去世后捐赠的藏品奠定了大英博物馆立馆之基）。大师大都不是那种优秀的学者，而是一些稀奇古怪之物的爱好者，会被这些东西吸引至咖啡馆一探究竟。1729 年，詹姆斯·索尔特在伦敦开了家"唐索尔特罗咖啡馆"，其中用来吸引顾客之物包括"来自耶路撒冷的彩色细带、救世主耶稣受鞭刑时被绑在上面的苦刑柱"，以及"一条 17 英尺长的巨蟒，从苏门答腊岛的鸽子窝里抓来的，肚子里还有 15 只家禽和 5 只鸽子"。①

咖啡馆的其他大主顾都是城市商人，他们会聚集在某些场所处理事务。其中最有名的便是劳埃德咖啡馆，成立于 1688 年，后来成为海上保险交易中心；还有堪称原始证券交易所的乔纳森咖啡馆，它在 1711 年的"南海泡沫"中起着举足轻重的作用，那场金融事件是由疯狂的股票投机引起的，随后导致市场崩溃。塞缪尔·皮普斯充分肯定了

---

① *A Catalogue of the Rarities to Be Seen in Don Saltero's Coffee House in Chelsea* (London, 1731).

咖啡馆作为联络网的价值，他在 1663 年决定要经常去咖啡馆而非酒馆，他喝咖啡喝到"几乎想吐"，但也因此从海军供应协议收取的回扣中获得了可观的财富。[①]

咖啡馆的成就与咖啡的成就不可混为一谈。早在 1660 年，咖啡馆里便可以选择热巧克力和茶作为替代饮品。1664 年，希腊人咖啡馆——新成立的英国皇家学会的摇篮——宣传本店不仅出售巧克力和茶，还教顾客如何制作这些饮品。颇能说明问题的是，1675 年，查理二世的大臣们再次试图打压咖啡馆，他们将任何出售"咖啡、热巧克力、冰冻果子露或茶"的场所都视为咖啡馆，但以失败告终。事实上，第一家热巧克力店"怀特家"直到 1693 年才成立，而托马斯·川宁开第一家茶店是在 1711 年，与其说这表明这些饮品直到后来才被接受，不如说这意味着人们更早之前便可接触到它们。

咖啡与咖啡馆之间的联系可能妨碍了家庭对它的接纳。咖啡馆基本是一个以男性为主的环境，鼓励人们与陌生人交流。唯一出现在咖啡馆的女性要么在上咖啡，要么在"提供（别的）服务"，都是为了满足顾客的需求。1674 年，《女性反对咖啡请愿书》严厉谴责了咖啡与咖啡馆，理由是它们使男人远离了家庭并变得性无能。请愿的倡议者大概是渴望夺回所流失的顾客的酒商们，他们有意利用了这种性别差异。

---

① Samuel Pepys, diary entry, Friday 23 January 1663, www.pepysdiary.com.

有教养的女性倾向于饮茶。茶受到了几位皇室模范成员的青睐，特别是葡萄牙布拉干萨王朝的凯瑟琳。1662年，她嫁给查理二世，并将茶引入英国王室。随后，英国两位女王玛丽二世（1689—1694年在位）和她的妹妹安妮（1702—1714年在位）都喜欢喝茶。女性可以聚在一起喝茶的地方要么是家里，要么是茶园这种公共场所，茶园的露天环境视野清晰，成为女士们聚会的体面场所。

想要估量伦敦咖啡馆火爆的程度并非易事。但下文中亨利·梅特兰提供的数字明显是精确的。1739年，他在首都伦敦进行了全面调查，发现共有551家咖啡馆——与人口数量的比值约为千分之一。光是金融城区有记录的咖啡馆就有144家——大致相当于酒馆和旅馆的数量。此外，梅特兰发现伦敦有8000多家杜松子酒馆，而在最穷的一个城区，这类酒馆的数量大约是该地区咖啡馆的80倍，这也就意味着咖啡是属于精英阶层的饮品。18世纪40年代以后，下层社会才开始饮茶。英国东印度公司进口中国茶的关税降低，但在国际市场上购买咖啡仍要承担很重的关税。

18世纪下半叶，咖啡馆在供应咖啡的同时开始提供酒饮，实际上又变回了酒馆，这从名字类似"土耳其人头像"的酒馆数量便能看出。例如，伦敦杰拉德街的"土耳其人头像"，该店曾在1764年组织了一个文学俱乐部，成员包括伟大的词典编纂者塞缪尔·约翰逊博士及其传记作者詹姆斯·博斯韦尔。两人分别习惯喝茶和葡萄酒。有些咖啡馆还兼作妓院，贺加斯在1738年的画作《早晨》中便有相关场景，他在画中描绘的是伦敦柯芬园的国王咖啡馆。那些进行商业活动的咖啡馆有时自行化身为交易所（如劳埃德咖啡馆），而大师会面的场所逐渐变成了绅士们的私人会所，从而维护其社会地位。1815 年出版的某部伦敦商业指南便只列了 12 家咖啡馆。

而与之形成鲜明对比的是法国咖啡馆。尽管法国咖啡馆起步较为缓慢，但在 18 世纪已发展成吸引所有阶层的社会场所。早在 17 世纪 40 年代，马赛便有了咖啡贸易，但直到 1669 年，咖啡才在巴黎为众人所知。同年，大概是法国驻君士坦丁堡大使为了给自己君主留下好印象，于是鼓动苏丹穆罕默德四世向路易十四派遣外交使团。使团在法国逗留了将近一年的时间，在被翻新成波斯宫殿的豪宅里，用咖啡等土耳其美食款待有影响力的朝臣们。这在法国社会的上流阶层中掀起了一阵"土耳其热"，正如莫里哀在《贵人迷》中所讽刺的那样。小贩们开始在圣日耳曼商业区的交易市场上售卖咖啡；1671 年，一位名叫帕斯卡尔的亚美尼亚人在巴黎开了第一家咖啡馆，但这股土耳其热潮退去后，它便倒闭了。

直到 18 世纪初，饮品商（酿酒商）、杂货商和药剂师行会之间关于咖啡销售权的长期争端才最终得到解决。饮品商获得了实际的垄断权，可以向坐在他们店里的顾客供应饮品，无论是柠檬水、咖啡还是杜松子酒。由此产生了咖啡馆这种形式的经营场所，顾客在咖啡馆里既可以喝咖啡，也可以喝酒，目的通常是利用前者的特性来中和后者的特性。

　　早期的一个例子是普罗可布咖啡馆。1686年，出生于意大利的饮品商弗朗切斯科·普罗科皮奥开设此咖啡馆，他将自己的名字改为弗朗索瓦·普罗可布。这家店的布置包括镀金镜子、大理石桌、顶部壁画以及枝形吊灯，让人想到的并非苏丹宫殿，而是贵族沙龙。咖啡及配餐都盛放在精美的瓷器中，配上银质餐具。普罗可布咖啡馆对面就是新成立的法兰西喜剧院，因此来自精英阶层的顾客可以在对周围环境感到安心的情况下在此邂逅戏剧演员们。在整个18世纪的欧洲，纵览诸如威尼斯的花神咖啡馆、罗马的希腊咖啡馆之类的气派咖啡馆，普罗可布咖啡馆堪称其中的典范。

画家不详，普罗可布咖啡馆，巴黎，约1700年。

到了 1720 年，巴黎大约有 280 家咖啡馆；1750 年，增加到了 1000 家左右；1790 年，又增至 1800 家，此时服务对象约有 65 万人。与普罗可布相比，这些咖啡馆大多迎合了更大众化的顾客的需求，为巴黎人提供了聚会与社交的场所，让他们可以在此玩跳棋等游戏，或赌各种各样的彩票。由于当时几乎到处都有人吸烟，因此这里早已备好陶土烟斗，装上烟丝，供顾客随时享用。咖啡馆在室内家具陈设上符合顾客的社会地位，同时也寻求额外的空间，于是延伸到了林荫道上。选址在郊区的咖啡馆租金较低，让顾客有机会白天享受田园风光，晚上寻欢作乐。

咖啡馆属于男性世界。尽管许多咖啡馆都是由伴侣共同经营——女人在前台工作，她的男性伴侣则在后厨"实验室"里准备饮品及配餐——但咖啡馆几乎没有其他女性踏足，因为这里本质上属于公共场所并且从事酒类贸易，她们担心自己会被误认为妓女。如果女性要喝咖啡，多半是让人将咖啡送到她们的马车内，以便私下饮用。

资产阶级女性之所以选择喝热巧克力，一个很重要的原因在于它所谓的药效。随着欧蕾咖啡①的流行，咖啡开始挑战热巧克力的首要地位。菲利普·迪富尔在他 1684 年出版的关于咖啡、茶和热巧克力的书中解释道："当（研磨好的）咖啡粉加入牛奶，煮至沸腾，稍稍变浓时，便会接近

---

① 欧蕾咖啡（ *café au lait* ），加入大量牛奶的咖啡。——译注

热巧克力的风味，几乎所有尝过的人都觉得不错。"①此外，欧蕾咖啡可以算是法式咖啡，而不是别国的咖啡。1690 年，一位对此极为热情的贵妇表示："我们这里既有好牛奶，也有好奶牛；我们想到可以撇去奶油……然后在牛奶中加糖和优质咖啡。"②

在低地国家③，不同性别和不同阶层的人对咖啡的接受更为迅速。在 18 世纪的阿姆斯特丹，下层与中产阶级家庭遗嘱清单上经常可见咖啡制作设备。早在 1726 年，便有人宣称咖啡"已经在我们的土地上基本发展开来，如今就连女仆和女裁缝早上都得喝点儿咖啡，否则无法穿针引线"。④在家中从事计件工作的织工们为了避免自己离开织布机，似乎也会来点儿加糖的咖啡让自己打起精神来。

咖啡的日渐流行促使欧洲各贸易公司设法确保其供应正常。1707 年，奥斯曼政府对帝国以外的地方实施咖啡出口禁令，供应变得紧张了起来。当时荷兰东印度公司的主管尼古拉斯·威森早在 1696 年便已开始在爪哇岛种植咖啡树。咖啡种子来自印度的马拉巴尔，当地传说最早种植咖啡树的是穆斯林学者巴巴布丹，他在前往麦加朝圣后，将

---

① Philippe Sylvestre Dufour, *Traitez nouveaux et curieux du café, du thé et du chocolat*, 3rd edn (The Hague, 1693), p. 135.

② Julia Landweber, 'Domesticating the Queen of Beans', *World History Bulletin*, XXVI/1 (2010), p. 11.

③ 低地国家（Low Countries），指欧洲西北沿海地区，通常指荷兰、比利时及卢森堡三国。——译注

④ Anne McCants, 'Poor Consumers as Global Consumers: The Diffusion of Tea and Coffee Drinking in the Eighteenth Century', *Economic History Review*, LXI, S1 (2008), p. 177.

种子藏在衣服里偷偷运了回来。对于马拉巴尔有咖啡，还有一种更说得通的解释——这是印度咖啡贸易的结果。

在爪哇岛，荷兰东印度公司强迫土著首领以事先定好的低价，提供固定量的咖啡。当地领主则以封建义务为由，要求其统治下的农户提供咖啡。对于农民而言，咖啡是一种既无法带来经济价值也无法带来营养价值的作物，因此他们缺乏提高种植技术的动力。为了完成配额，他们更倾向于在自家土地或林地种植咖啡树。爪哇岛的种植业主要集中在西部，当地农户有时会被迫迁至合适的咖啡种植园，由领主的手下管理。①

1711 年，从爪哇岛到荷兰开通了定期航运，促使阿姆斯特丹成立了欧洲第一家咖啡交易所。1721 年，阿姆斯特丹市场上 90% 的咖啡都来自也门；而到了 1726 年，则 90% 都来自爪哇岛。②直到 18 世纪中叶，该岛的供应量一直不断增长，但随着加勒比海地区新种植园的出现，数量又逐渐减少。

荷兰人对此负有一定的责任。他们在 1712 年将咖啡引入苏里南，那是位于拉丁美洲大陆东北海岸线上的一块殖民飞地，濒临加勒比海。苏里南的咖啡出口贸易始于 1721 年，到 18 世纪 40 年代出口量便超过爪哇岛。在苏里南，种植

① M. R. Fernando, 'Coffee Cultivation in Java', in *The Global Coffee Economy in Africa, Asia and Latin America, 1500–1989*, ed. William Gervase Clarence Smith and Steven Topik (Cambridge, 2003), pp. 157–72.
② Steven Topik, 'The Integration of the World Coffee Market', in *The Global Coffee Economy*, p. 28.

者除了种植咖啡树外别无选择——而真正种植这种作物的是种植园里的奴工。

苏里南利弗普尔咖啡种植园面貌，约1700—1800年。这幅出色的匿名画展示了该咖啡种植园的规模。主建筑紧挨登陆区和水闸后方。在底图中，左边是三面围绕的奴隶居住区，背后则是一小块一小块的配给土地。右边是处理区，从图上可以看到工人们正在露天空地上耙咖啡豆，将之晒在高台上。种植园主管与监工则住在中间偏右的"白色"居住区，而他们的休闲花园则分布在最右边。

1715 年，法国人在波旁岛（今留尼汪岛）种植咖啡树，该岛属于非洲东海岸马斯克林群岛的一部分。17 世纪 40 年代，法国东印度公司开始殖民统治这个无人居住的荒岛，并向法国移居者授予土地特许权，而在地里干活的都是非洲奴隶。事实证明，阿拉比卡咖啡树虽然是无视奥斯曼帝国禁令从也门引进的，但在此地的种植非常成功，因此法国东印度公司于 1724 年颁布法令，将收回过去所有不种植咖啡的土地的特许权，甚至还讨论过要对故意破坏咖啡树者处以死刑。①

18 世纪 20 年代，法国人还将咖啡引进其加勒比海地区的殖民地，首先是马提尼克岛。据传，年轻的海军军官加布里埃尔·德克利厄 1723 年从巴黎的植物园将咖啡树插枝运到了该岛，而多年后，他公开讲述了自己在航行中如何将定量分配给他的水分给这些插枝的英勇事迹。现在看来，该岛在 1724 年便开始使用来自波旁岛和苏里南的咖啡种子种植咖啡树。

咖啡树被带到圣多明各（今海地）的具体时间及方式更为模糊，但咖啡树在此地成功种植，产量很快就超过了加勒比海地区其他地方。法国在 1697 年"九年战争"②结束时获得了该殖民地，当时伊斯帕尼奥拉岛被一分为二。

① Gwyn Campbell, 'The Origins and Development of Coffee Production in Réunion and Madagascar', in *The Global Coffee Economy,* p. 68.
② 九年战争：又称大同盟战争，战争缘起是法国国王路易十四欲在欧洲进行大规模扩张，遭到了荷兰、神圣罗马帝国哈布斯堡王朝、英国、西班牙等国组成的大同盟联合对抗。——译注

东部地区（今多米尼加共和国）仍归西班牙统治，而圣多明各则占据了该岛三分之一的西部山地地区。与加勒比海地区其他地方一样，沿海地势较低地区主要种植甘蔗，而内陆山区则建咖啡农场。

18世纪30年代以前，法国东印度公司一直不允许波旁岛或加勒比海地区收获的咖啡在法国本土销售。此举是为了保护自己对价格更高的摩卡咖啡的垄断权。于是，这些产地的咖啡反而被运往阿姆斯特丹交易所——其中包括马龙咖啡，如今被称为毛里求斯咖啡，它是在波旁岛上发现的野生咖啡品种。事实证明，它的品质不如人工种植的阿拉比卡咖啡，因此在18世纪20年代便不再种植。到了18世纪50年代，美洲咖啡在阿姆斯特丹交易所的交易量与亚洲咖啡差不多。

等到17世纪中叶[①]咖啡进口禁令解除后，产自殖民地的咖啡纷纷涌入法国，导致咖啡价格下降，从而使下层社会更容易享用到这种饮品。而围绕咖啡的饮用方式，社会上形成了一定程度的势利之风，哲学家雅克-弗朗索瓦·德马希1775年做过一场比较。

> 一位是来自上流社会的女性，惬意地坐在扶手椅上，享用着汁多味美的早餐，漆饰茶桌明亮光滑，四周弥漫着摩卡咖啡的香气，盛放咖啡的是……

---

① 此处疑原文有误，根据上下文，应为18世纪中叶。——译注

V'la la Mᵈᵉ de Café au Lait.

à Paris, chez L. M. Petit, Mᵈ d'Estampes rue Sᵗ Martin Nº 96. au grand Raphaël
chez Martinet, rue du Coq, Nº 13 et 15.
et rue des Mathurins, Nº 18.

Dép²ᵉ à la Bᵗᵏᵉ qᵘᵉ Impᵗᵉ

《一位卖欧蕾咖啡的女人》，出自阿德里安·若利的《巴黎的艺术、手工艺品与街头小贩》系列画（1826年版）。

镀金瓷杯，咖啡中加入精制糖和优质奶油；而……另一位是卖菜的女人，她将馊掉的便士面包浸泡在一种令人作呕的液体中，有人告诉她那是欧蕾咖啡，不过是装在简陋陶壶里的。[①]

到了 18 世纪 80 年代，全球 80% 的咖啡供应都来自加勒比海地区，主要是圣多明各。18 世纪 60 至 80 年代，该地种植园数量不断增长，直到咖啡出口总值与甘蔗不相上下。这个殖民地成功经营下去的原因在于生产成本低——主要通过进口非洲奴隶作为劳动力。

1798 年，种植园主 P. J. 拉博里出版了一本关于圣多明各咖啡种植的指南，内容引人入胜但同时也令人不安。他描述了咖啡种植从开垦土地到豆子装袋的所有阶段。书中还介绍了"西印度群岛式处理法"，西印度群岛人创造性地进行水洗脱胶，即利用水槽系统软化果肉，便于之后进行一系列的脱胶处理。然而，值得注意的是，他坚信"黑鬼"（他对奴隶的称呼）"这种生物天生就是奴隶，需要我们强制性地使之保持这种状态，才能让其提供必要的服务；因为……要是换种情况，他们便不会劳动了"[②]。

在买奴隶时，拉博里还列出了建议挑选的特征，如看上去开朗大方，眼神干净有神采，牙齿健全，手臂强壮有力，双手大而干燥，腰臀健硕，以及四肢灵活协调。奴隶被买

---

① Emma Spary, *Eating the Enlightenment* (Chicago, IL, 2012), p. 91.

② P. J. Laborie, *The Coffee Planter of Saint Domingo* (London, 1798), p.158.

下后，被强迫服用"发汗剂"两周，目的是通过出汗消除航行途中染上的疾病，然后给他们打上烙印，此行为"讨人厌却有必要"。

新奴隶必须"被调教"——他们要一边适应更为寒冷的气候，一边逐步学习干农活。拉博里倾向于买15岁左右的少男少女，他们在从事园艺和除草的同时，可以按照"主人的想法"进行培养。接着，他们将会加入种植的主力军，从早到晚地在种植园里干活，受监工——一位由主人委托、手握鞭子的奴隶——"驱策"。

维护权威最为紧要。与奴隶间犯下的任何罪行（包括暴力攻击和强奸）相比，不服从命令（如顶撞主人或监工）受到的惩罚会更为严厉。拉博里在书中写道，两次鞭刑之间需要清洁鞭子，避免传染。

圣多明各的种族政治颇为复杂。所谓的"有色自由民"拥有超过三分之一咖啡种植园以及四分之一的奴隶。这群人中既有经父辈承认的法国殖民者的混血后代，也有越来越多的被主人释放的前黑奴。到了1789年，这块殖民地上有2.8万名"有色自由民"和3万名白人，而这两个群体在数量上都不敌奴隶，其数量为4.65万。

受1789年法国大革命鼓舞，"有色自由民"主张自己拥有与白人平等的权利，而奴隶们为了改善条件，则利用不稳定的局势发动叛乱。自1791年起，几股势力结成了一个并不稳定的联盟，由被解放的黑奴杜桑·卢维杜尔领导，他曾一度拥有一座咖啡种植园和15名奴隶。可怕的是，暴力、

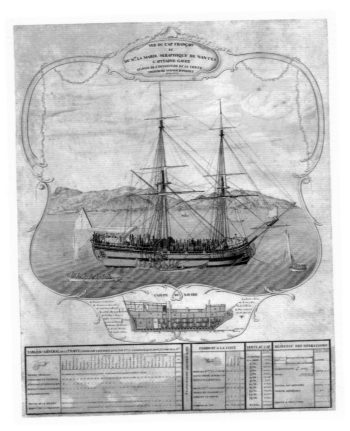

《"拉玛丽-塞拉菲克号"》，1773 年。这幅匿名水彩画展现了这艘在南特注册的法国奴隶船从安哥拉出发，最后停泊在圣多明各法兰西角的场景。画中是船上的活人被当作货物贩卖的第一天。整艘船由一排铁栅栏一分为二，奴隶被关在主甲板上，而拍卖在后甲板上进行，欧洲买家则在船尾享用野餐。

外国干涉、压制和战争在此轮番上演，一直持续到1804年圣多明各宣布独立，更名为海地，并废除了奴隶制。一千多座咖啡种植园也因此遭到摧毁，其中包括拉博里的那座，当时他已逃到牙买加。虽然这里后来又建立了新的农场，但实际已经丢掉了原本的咖啡贸易，因为欧洲各国和美国由于害怕黑人统治合法化而选择避开海地。

而英国海军对法国领土的封锁，进一步导致欧洲内部的咖啡供应中断。拿破仑对此采取的措施是鼓励采用本土菊苣作为替代品。普鲁士的腓特烈大帝也提倡使用菊苣制作饮品。为了打压咖啡消费，他在18世纪80年代还雇用了一批所谓的"咖啡嗅探军"[1]。当时普遍的做法是在咖啡豆中加入烤过的菊苣来增加其体量。甚至在20世纪初，《茶与咖啡》贸易月刊的创始编辑威廉·尤克斯还抱怨，许多欧洲人已经"逐渐习惯了菊苣和咖啡混合的味道，因此如果他们有机会喝到一杯真正的咖啡，能否喜欢还真是难说呢"[2]。

尽管如此，整个欧洲的咖啡消费在19世纪上半叶持续增长。这一时期瑞典小说的特色便是描写社会各个阶层都在喝咖啡的场景：例如，在埃米莉·弗吕加勒-卡伦1844年出版的小说《波尔·韦尔宁》中，男主人公是个贫穷的渔夫，为了给生病的母亲买咖啡而踏上了一趟危险的旅程。在既经营酒类生意又卖其他商品的店里，他邂逅了一位上了年

---

[1] 即嗅出非法烘焙咖啡豆的人。——译注

[2] W. H. Ukers, *All About Coffee* (New York, 1935), p. 554.

纪的女服务员，她"就坐在厨灶旁，嘴里叼着烟斗，咖啡壶就放在火炉上……真是在享受人生极乐啊"。当时，咖啡开始在名为"殖民商品"的店里销售——这个名字实在贴切，因为欧洲的咖啡供应仍大多来自帝国的殖民地。

首个滴漏式咖啡制作设备诞生于19世纪初，即所谓的德贝卢瓦伊壶，得名于热爱咖啡的巴黎大主教。在上下隔开的过滤层上放入咖啡粉，让热水从上层过滤到下层。后来的咖啡壶设计可以让水在炉子上加热，然后将整个设备倒置，便可进行过滤。[①]在整个19世纪，虹吸管系统和流体静压渗滤壶等时尚设备颇受精英阶层青睐，而欧洲大部分地区普遍使用的还是滴漏式设备。

圣多明各咖啡贸易的消亡点燃了亚洲咖啡生产的复兴之火。爪哇岛咖啡的盛行使这座岛的名字在美国成为咖啡的代名词。然而，以"爪哇"之名出售的咖啡可能源自印度尼西亚群岛的苏门答腊岛等荷兰殖民地。当时咖啡从此地运往纽约可能需要五个月，其间咖啡豆会逐渐老化，并经常因为水分渗出而变成棕色。这种咖啡由于酸度低而格外珍贵，人们甚至在蒸汽船出现后仍继续用帆船运输。

荷兰殖民当局继续借助当地统治者进行管理，引入所谓的"征收制度"，要求农户留出自家部分土地或劳动力来种植经济作物，并且专门销售到荷兰。1860年，一位前行政官员写了本自传小说《马克斯·哈维拉尔》，书中展现了

---

① Enrico Maltoni and Mauro Carli, *Coffeemakers* (Rimini, 2013).

荷兰人迁就着贫穷的领主时，农民又是如何挨饿的[1]。到了19世纪80年代，爪哇岛60%的农户都被迫种植咖啡树。照料这些咖啡树占用了他们15%的时间，但由于要按固定的低价出售，因此咖啡给他们带来的收入只占其总收入的4%。

创作者：科菲，约19世纪70年代。该版画由位于荷属东印度群岛首府巴达维亚（今印度尼西亚首都雅加达）的科尔夫出版公司制作。画中展现了咖啡采摘、处理和制作的各个阶段，尽管这些殖民地的人民，尤其是本土家庭，几乎不会消费当地产的咖啡。画上看到的咖啡壶属于德贝卢瓦伊壶中的一款，热水由上壶注入，通过一层咖啡粉渗至下壶。画中出现的牛奶罐表明此海报是针对荷兰消费者设计的。

---

① Multatuli, *Max Havelaar: Or the Coffee Auctions of a Dutch Trading Company* [1860] (London, 1987).

英国人也扩大了其殖民地的咖啡生产，尤其是在锡兰（今斯里兰卡）。他们在拿破仑战争期间从荷兰人手中夺取了该海岸地区的控制权，并且开始征服内陆，于1815年推翻了独立的康提王国。英国前来开拓的企业家们为了建立咖啡种植园，在此砍伐森林，猎杀了岛上的许多大象，还从印度马德拉斯地区负债累累的泰米尔人中进口劳动力。在前往这些种植园的途中，或是由于当地的工作条件，总之有无数的人丧命。[1]

到了19世纪60年代末，英国在锡兰和印度两地的咖

咖啡叶锈病。叶锈病斑的迹象最早出现在叶背面。

---

① Donovan Moldrich, *Bitter Berry Bondage: The Nineteenth Century Coffee Workers of Sri Lanka* (Pelawatta, Sri Lanka, 2016).

啡总产量接近荷兰殖民地。1869 年，暴发了一场由咖啡驼孢锈菌引起的叶锈病。等到 19 世纪 80 年代中期，两地的咖啡种植园大多已被毁，于是转变成了茶叶种植园，也因此巩固了茶叶在英国胜过咖啡豆的地位。而到了 1913 年，锡兰已成为咖啡净进口国。

叶锈病蔓延到整个亚洲，摧毁了爪哇岛、苏门答腊岛、东印度群岛其他地区，以及印度的大部分咖啡生产，甚至还波及非洲和太平洋群岛。一些种植者以利比里亚本土品种大果咖啡作为阿拉比卡咖啡的替代品。这个品种的咖啡豆味涩，几乎只有马来西亚和菲律宾当地人喜爱，在菲律宾还成了深度烘焙、咖啡因含量高的巴拉科咖啡的主要原料。无论怎样，大果咖啡也被证明易受此锈菌侵害。直到第一次世界大战爆发，亚洲的咖啡供应量也只占全世界的二十分之一，而在叶锈病暴发前，其供应量大约占到了三分之一。当时全球咖啡经济主要以美洲为中心。

温斯洛·霍默，《咖啡的召唤》，1863 年。在这幅版画中，波托马克军团的士兵们正排队等候架在篝火上的大桶里煮的咖啡。

工业产品

19世纪下半叶，巴西和美国两个美洲国家将咖啡转变为一种工业产品。在不大幅提价的条件下，巴西迅速扩大咖啡产量，使美国将之纳入自身不断扩张的消费经济。通过从欧洲引进农民工来代替奴隶作为劳动力，巴西将咖啡树种植的疆域扩展到了腹地。19世纪中叶至20世纪中叶，由于咖啡消费者从家庭烘焙转向购买预先加工好的、有品牌的工业咖啡产品，美国人均咖啡消费量增加了两倍。中美洲和哥伦比亚开始竞争美国市场，各国都在竭力保护本国利益，于是便出现了新形式的咖啡政治。

美国咖啡：从殖民时代到内战时期

美国人之所以喜欢咖啡，常被认为是那场争取独立的斗争的结果。对于美国殖民地居民"无代表，不纳税"的要求而言，茶叶成了具有象征意义的焦点。此前，英国政府一直对进口到美国殖民地的茶叶征收关税，这也是东印度公司垄断贸易的一部分。1773年12月16日，抗议者发动了"波士顿倾茶事件"，将一箱箱茶叶从切萨皮克湾港口的船上倾倒入海。从此，便开始流传美国爱国者们改喝咖啡的故事。

1880—1950 年美国咖啡消费统计数据[①]

年份	总进口量 （百万磅）	人均消费量 （磅）	占世界进口量的 比重（%）
1800	8.8	1.65	
1830	38.3	2.98	
1860	182.0	5.78	28.7
1890	490.1	8.31	36.1
1920	1244.9	11.88	56.1
1950	2427.7	16.04	63.6

现实更为复杂。咖啡在美国各个殖民地早已有之；尤其是在波士顿，1670 年，多萝西·琼斯成为此地首位特许出售"咖啡和热巧克力"的商人。咖啡馆由此在整个波士顿扩张开来，大多兼作酒馆，其中 1697 年开张的绿龙咖啡馆便是政治活动家的定期聚会场所。这些殖民地的咖啡虽然主要从牙买加进口，但也受英国控制。

"倾茶事件"后，这些爱国人士选择获取英国供应物的替代品。1774 年，约翰·亚当斯[②]要求"茶叶确实走私而来或没有缴纳关税"。1777 年，在妻子阿比盖尔描述了波士顿妇女如何闯入仓库，寻找咖啡和糖的情况后，亚当斯希望"女性摆脱对咖啡的依恋"，开始饮用由美国本土产品制成

---

① Derived from William H. Ukers, *All About Coffee* (New York, 1935), p. 529; Mario Samper, 'Appendix: Historical Statistics of Coffee Production and Trade from 1700 to 1960', in *The Global Coffee Economy in Africa, Asia, and Latin America, 1500–1989*, ed. William Gervase Clarence Smith and Steven Topik (Cambridge, 2003), pp. 419, 442–4.
② 约翰·亚当斯（John Adams，1735—1826），美国独立运动领导人之一，美国第二任总统。——译注

的饮料。[1]

新独立的美国开始获得法国殖民地圣多明各的咖啡供应后，咖啡在美国普及开来。到了 1800 年，在美国，咖啡的人均消费量超过了 680 克（约合 1.5 磅）。[2]在拿破仑战争期间，美国扩大了利润丰厚的咖啡再出口贸易。为防止海军封锁，来自加勒比海地区的咖啡是用美国船只运往欧洲的。

1820 年后，咖啡的消费量显著增加，因为价格下跌了：从 1821 年的 21 美分 / 磅跌至 1830 年的 8 美分 / 磅。价格下跌是因为投机者预测法国与西班牙之间会爆发战争，于是大量囤积咖啡，结果战争并没有爆发，他们只好去国际市场上倾销。随着世界咖啡供应量的扩大，咖啡价格在接下来的 20 年里很少超过 10 美分 / 磅。1832 年，美国联邦政府取消了咖啡进口税，到了 1850 年，咖啡的人均消费量超过 2.3 千克（约合 5 磅）。

在圣多明各咖啡供应中断后，古巴成为美国的主要供应国，许多美国投资者都在该岛购买了种植园。但 19 世纪 40 年代的一系列自然灾害摧毁了古巴成百上千棵咖啡树，许多投资商因此转而种植甘蔗。此后，美国逐渐从拉丁美洲，尤其是巴西进口价格低廉的咖啡。

美国内战（1860—1865 年）在美国咖啡史上至关重要。联邦军队就没有中断过喝咖啡：每天大约 43 克（约合 1.5

① Steven Topik and Michelle McDonald, 'Why Americans Drink Coffee', in *Coffee: A Comprehensive Guide to the Bean, the Beverage and the Industry*, ed. R. Thurston, J. Morris and S. Steiman (Lanham, MD, 2013), p. 236.
② Figures derived from William H. Ukers, *All About Coffee*, p. 529.

盎司）的配给量，一年的总量竟然达到 16 千克（约合 36磅）。这个量可以轻易满足他们每天喝上十杯咖啡。作战的将军们意识到咖啡因具备提神的功效，于是战斗前都要确保手下士兵已喝了大量咖啡；一些士兵将咖啡研磨机直接安装到自己的枪托上。联邦军队封锁了南方海岸线，这意味着南方各州——以及他们的部队——被迫用菊苣和橡子等植物作为咖啡的替代品。

"麦金利送咖啡之旅"。安提塔姆国家战场的纪念碑详图，描绘了麦金利（后来成为美国总统）在内战期间送咖啡的壮举。他在 1901 年被暗杀，该纪念碑建于 1903 年。

那个时期，在士兵的日记里，"咖啡"一词出现的频率比"步枪""大炮"或"子弹"都高，由此可见咖啡对于当时部队生存的重要性。分发定量咖啡的军士在叫士兵领取个人的配给量时，为避免被人指责分配时偏心，还会特意将身子转向另一边。炮兵约翰·比林斯在回忆录《压缩饼干与咖啡》中描写了如何通过将压缩饼干泡进咖啡来杀死饼干中的象鼻虫，死掉的虫子会浮在咖啡表面上。他回忆道：

> 如果下令午夜行军……必须先喝一壶咖啡；如果下令在中午十点左右或下午停战，咖啡同样必不可少……用餐时以及两餐之间都要来点咖啡……如今，还能受得了这么喝咖啡的老兵得是全军咖啡瘾最大的那批人。[1]

1862 年 9 月 17 日在安提塔姆的那场战役见证了整个内战期间最血腥的一天。19 岁的威廉·麦金利中士（后来成为美国总统）冒着猛烈的炮火，沿着前线一路为部队提供咖啡。据他们的指挥官讲，咖啡对士气的作用"好比让一个新的团参加战斗"[2]。

---

① John D. Billings, *Hardtack and Coffee* (Boston, MA, 1887), pp. 129–30.
② Jon Grinspan, 'How Coffee Fueled the Civil War', www.nytimes.com, 9 July 2014.

咖啡业的基础

内战结束后，士兵回到家乡，他们喝咖啡的习惯刺激了国内新兴的咖啡业。到 19 世纪 80 年代，美国咖啡进口量占世界总量的三分之一，并于 1882 年促成了纽约咖啡交易所的诞生。

整个 19 世纪，美国农村大部分地区主要从供应商目录或杂货店大批量购买咖啡生豆。然后购买者自己在家将一批批咖啡豆倒入平底锅，用柴火炉烘焙，来回翻炒大约 20 分钟。经济条件稍好的家庭可能拥有密封的家用烘焙炉，可以手动或以蒸汽翻炒。到了 19 世纪中叶，虽然家用咖啡研磨机日渐普及，但在将烤好的咖啡豆磨成粉时依旧经常用到杵臼。

制作咖啡的方法很简单——只需将磨好的咖啡与水放入壶中加热。家庭指南建议加热至沸腾后，煮 20 至 25 分钟。要使咖啡粉沉淀到壶底，需使用各种各样的添加剂——其中最常见的便是蛋清，当然还有鱼胶（一种以鱼类胶原蛋白做成的明胶）。

最早为大众接受的咖啡萃取设备是 1859 年的欧道明咖啡壶。它属于早期的咖啡渗滤壶，使用原理是将咖啡倒入这种多孔的容器内，底部盛水煮至沸腾，同时用上方的冷凝器将底部冒出的水汽液化并回收。建议使用者将壶中的咖啡和水在炉上放置一夜，次日早餐前再煮 10 至 15 分钟。由此制作出来的咖啡口感薄而味道偏苦，这便成了美式咖啡的特色口味。

到了 19 世纪 40 年代，出现了一批主要城市，为新的咖啡烘焙批发业务创造了条件。烘焙批发商按重量向商店供应现烘好的散装咖啡豆。这些咖啡豆经由伸缩式烘焙炉烘焙，1846 年，来自波士顿的詹姆斯·W. 卡特获得了伸缩式烘焙炉的专利权。这种烘焙炉有一个长长的烘焙筒，可放进烧煤的砖炉内。烘焙筒借由滑轮系统在炉内拉进拉出，并通过其两侧的滑动门进行填充与倒出。

早期工业化咖啡烘焙厂中正在运作的卡特伸缩式烘焙炉。每个烘焙筒可装大约 90 千克（约合 200 磅）咖啡豆。图为弗朗西斯·瑟伯所著《咖啡：从种植园到咖啡杯》（1887）一书中的原版版画。

烘焙者可根据筒边冒出的烟雾颜色来判断咖啡豆是否烘焙完成。他们把冒着热气的咖啡豆倒入托盘内，然后用手翻搅这些豆子，直至冷却。有些人干脆把这些热气腾腾的咖啡豆倒在地板上，用耙子将之铺开并洒上水。一位观察者回忆道："水一接触到热气腾腾的咖啡豆，便会出现大量蒸汽，因此每当一批咖啡豆从炉火中抽出后，烘焙室有好几分钟都会笼罩在浓雾中。"[1]

1864年，杰贝兹·伯恩斯获得了他发明的自排式烘焙炉的专利权。该烘焙炉的形态是安装在砖炉内的旋转烘焙筒，咖啡豆借由所谓的"双螺杆"均匀地上下移动，从而达到均匀的烘焙效果。该设备运作的关键在于可以将咖啡豆从筒前倒入冷却盘中，而无须将整个烘焙筒从炉火中取出。伯恩斯还改进了咖啡豆的冷却和研磨技术，有省时之效，从而大大缩小了咖啡生豆和烘焙好的咖啡豆在批发和零售方面的差价。1874年，伯恩斯宣布：

> 如果可以买到处理适当的咖啡豆，那么家庭烘焙在本国任何地方都会持久存在，这种说法简直荒谬……若大烘焙厂可以好好烘焙，那么就不会有人为小店烘焙买单……烘焙时适当注意，大工厂不仅可以……为自己争取到高销量，还可以控制其他方的烘焙。[2]

在当时的美国，咖啡势必成为一种大规模生产的工业产品——面向新兴的消费社会进行品牌化和营销。

---

[1] Ukers, *All About Coffee*, p. 589.
[2] 同上，p. 589。

咖啡品牌的崛起

与兄弟在匹兹堡经营杂货店批发业务的约翰·阿巴克尔属于新一代伯恩斯烘焙炉的首批购买者之一。1865年，他开始将烘焙过的咖啡豆装在（原本用于装花生的）加厚包装袋里出售。三年后,他为一种用于烘焙的蛋糕釉申请专利，称它可以使咖啡豆表面隔绝空气并起到净化的作用，从而防止咖啡豆变质。在他的广告上，一位妇女在柴火炉前烘焙咖啡豆时抱怨："唉，我的咖啡豆又烤焦了。"而她衣着讲究的客人此时建议她："购买阿巴克尔兄弟的烘焙咖啡豆，像我一样，远离烦恼。"下面的文字是："你自己并不能恰到好处地烘焙咖啡豆。"[1]

1873年，阿巴克尔兄弟推出了阿里奥萨咖啡，成为美国首个全国知名咖啡品牌：一粒粒光滑的咖啡豆被装进独特的黄色包装袋中，上面印有"Arbuckles"（阿巴克尔兄弟）的红色字样，而"阿里奥萨"品牌名上方的商标设计图案是个飞翔的天使。到了1881年，该公司用纽约和匹兹堡工厂里的85台伯恩斯烘焙炉进行豆子烘焙，并在芝加哥和堪萨斯州设有分销库。

阿里奥萨牌咖啡最专注的是美国西部由牛仔、牧场主和拓荒者组成的市场。他们中许多人都是参加过内战的退伍

---

[1] Mark Prendergast, *Uncommon Grounds* (New York, 2010), p. 49.

军人，早已爱上了喝咖啡。每包咖啡都有一根薄荷棒，目的
是用薄荷的甜味中和咖啡的味道。据说，大篷车队[①]的厨师
会以"谁想吃糖？"的呼唤来吸引人们自愿磨豆子。每包
都有优惠券，可用来兑换工具、枪支、剃须刀、窗帘甚至
婚戒等物品。天使形象的商标设计是用来使美国原住民相信
咖啡可以赋予其精神力量，让人体验到咖啡因所带来的兴
奋感。

阿巴克尔的阿里奥萨咖啡贸易名片

烘焙咖啡批发贸易的兴起带动了其他几个著名咖啡品
牌的发展。在 19 世纪 50 年代的"淘金热"期间，吉姆·福
尔杰在旧金山成立了福尔杰咖啡烘焙公司。1878 年，凯莱
布·蔡斯和詹姆斯·桑伯恩将他们最初在波士顿经营的咖啡

————————
① 大篷车队（wagon train），指 19 世纪美国人向西部迁移时用来运输
的马车队。——译注

公司合并，创立了海豹牌①咖啡——第一个使用密封罐包装的品牌。

罐装咖啡成了美国咖啡包装的标准，然而在这一包装过程中也封进了空气，因此咖啡变质问题仍然有待解决。1900年，旧金山另一家公司希尔斯兄弟推出了真空包装的咖啡，解决了这一问题。这项技术倾向于选择预磨咖啡，如希尔斯兄弟公司的高端品牌红罐咖啡。1892年，奇克－尼尔公司推出了麦斯威尔——得名于纳什维尔一家豪华酒店，它们向该酒店供应咖啡。

到了1915年，85%的消费者都更愿意购买预先包装好的品牌咖啡，而不是散装的烘焙咖啡豆。当时大约有3500个咖啡品牌，不过并非所有牌子的咖啡都摆放在当地杂货店的货架上。选择快递公司送货上门的消费者占市场的60%左右：当时最大的配送公司——宝石茶公司②有一半的收入来自咖啡销售。另一个占据大量市场份额的是连锁商店自己品牌的咖啡。大西洋与太平洋茶叶公司，通常被称为A&P，推出自己的品牌：八点钟咖啡，公司还通过在店里安装咖啡研磨机来增添一种"戏剧效果"。

20世纪初，咖啡的消费量为人均5千克（约合11磅），巩固了咖啡作为美国国家饮品的地位。如今，美国咖啡的进口量超过世界一半的供应量，烘焙公司用"水牛"和"餐

① 英文为seal，兼有"海豹"和"密封"之意。——译注
② 此处疑作者笔误，原文为Jewell Tea Company，经查证应为Jewel Tea Company。——译注

车特供"等名字将自己定位为美国固有品牌。托马斯·伍德公司吹嘘自家的山姆大叔咖啡产于"自己在波多黎各、夏威夷和马尼拉的咖啡种植园"（原话如此）。

蔡斯与桑伯恩咖啡公司旗下的"海豹牌爪哇摩卡咖啡"——首批罐装咖啡。随着咖啡原产地声明的规定愈发严格，该品牌名称被缩写为"海豹"。

不过，大多数烘焙公司都不会透露自家混合咖啡的原产地。1897 年，希尔斯兄弟公司以一位阿拉伯人身着飘逸长袍的形象申请了商标，利用"大篷车""桑托拉""蒂明戈""撒克逊人"等品牌名来掩盖所生产咖啡的更多信息。只有爪哇与摩卡仍然是公认的咖啡产地，传播了牛仔称呼咖啡的俚语"爪摩卡"。阿巴克尔兄弟提醒消费者"谨慎购买劣质包装的咖啡，商家通常会谎称它们是由来自摩卡、爪哇和里约的咖啡豆混合而成的，此乃制造商为欺骗粗心大意的顾客而使用的低劣花招[1]"。人们普遍认为阿里奥萨咖啡由来自里约和桑托斯的咖啡豆混合而成。到了 19 世纪 70 年代中期，美国消费的咖啡 75% 以上都来自巴西。

巴西咖啡

据传，咖啡是由弗朗西斯科·德·梅洛·帕列塔在 1727 年引入葡萄牙殖民地的。故事的经过是，巴西[2]外交官帕列塔被派去解决荷兰和法国在圭亚那殖民地的纠纷。回国时，他带着情人（法国总督夫人）送给他的一束花，里面藏着咖啡种子。他在家乡帕拉地区种下了这些种子。但直到 1822 年，与糖相比，咖啡在巴西的产量仍然较低。

圣多明各咖啡供应中断后，咖啡价格上涨，里约热内卢南部的帕拉伊巴山谷地区因此开始种植咖啡树，从而改变了咖啡的运势。咖啡树适宜红紫土种植，这种排水良好、营养丰富的红黏土在巴西中南部各州均有分布。

当时的栽培技术原始粗暴，几乎无视环境问题。山坡上的树林直接被砍伐焚毁，在土壤上方形成了一层肥沃的灰，咖啡树幼苗就栽于其中。整个种植过程丝毫没有考虑水土流失的问题，咖啡树在充足的阳光下生长，吸走了土壤的养分。咖啡产量提高，也是因为更多未开垦的土地被用于种植咖啡树。

---

① Mark Prendergast, *Uncommon Grounds* (New York, 2010), p. 71.
② 巴西在 16 至 19 世纪初为葡萄牙殖民地。——译注

1870—1990 年巴西咖啡产量统计[1]

年份（取两年的平均数）	巴西产量（百万袋）	世界产量（百万袋）	占世界产量份额（%）
1870—1871	3.1	6.6	46.9
1900—1904	14.5	18.7	77.5
1930—1931	25.1	37.0	67.8
1960—1961	32.9	68.9	47.7
1990—1991	28.5	98.4	28.9

富有的上层人士拥有大种植园，让奴隶种植咖啡树。每个奴隶大概要照料 4000 至 7000 株咖啡树。中间几乎不怎么养护。咖啡豆经自然干燥处理完，便直接脱壳，然后用骡车运往里约。由于缺少土壤养护，里约咖啡豆的声誉并不怎么好，容易发霉并产生异味。如今仍用"里约味"来形容具有这类瑕疵的咖啡豆。

在 1807 年美国禁止进口奴隶后，北美的奴隶贩子便转向巴西市场，建立三角贸易：以美国商品换取非洲奴隶，然后将这些奴隶卖给巴西，并在此购买咖啡再运回美国。这场大西洋奴隶贸易直到 1850 年英国通过海军直接干预才结束。

当时，现有的奴隶（约占人口三分之一）仍是巴西经济的核心。巴西南部的咖啡种植者会从北部购买奴隶，如此形成了一个巴西内部的奴隶贸易市场。直到 1871 年，巴西才通过了所谓的"自由子宫法"，赋予奴隶所生子女自由，随后在 1888 年通过了"黄金法"，释放了剩余所有奴隶。

---

[1]　Data from Francisco Vidal Luna and Herbert S. Klein, *The Economic and Social History of Brazil since 1889* (Cambridge, 2014), pp. 355–9.

描绘奴隶搬运咖啡豆的水彩画，根据让-巴普蒂斯特·德布雷1834年所著的《巴西之旅》创作。领头者通过弹奏拇指琴来控制队伍步调。

1872年[①]，巴西帝国统治被推翻。一个新的共和国成立，保利斯塔——圣保罗州的咖啡巨头们——占据主导地位。

圣保罗地区的主导地位

保利斯塔以贫穷的欧洲移民取代奴隶作为劳动力。这些移民在种植咖啡树的大型庄园通过劳力赚钱，种植园提供住宿和一小块土地，供其自行种植食物。1884年，巴西政府开始出资补贴运输移民的初始费用，于是到了1903年，

① 此处原文疑有误，经查证应为1889年。——译注

此地已有 200 多万名移民。其中超过一半的人来自意大利，他们被许诺的土地吸引而来，但最终发现自己实际成了契约劳工，需要以劳动偿还被运输而来的路费。这些劳动条款相当苛刻，因此 1902 年意大利政府便禁止有补贴的移民计划。于是，葡萄牙和西班牙随后成为巴西契约劳工移民的主要来源地。

当时，咖啡产量大幅上升——从 1890 年的 550 万袋增长到了 1901 年的 1630 万袋。1901 至 1905 年，巴西咖啡

20 世纪 30 年代的巴西桑托斯港。工人们正在卸下一袋袋咖啡，把它们扔进地下滑道中，这样便直接送进了船舱里。

产量占世界总产量的 73%。巴西咖啡树绝大部分都种植在圣保罗地区，而到了 1900 年，那里种植的咖啡树数量已超过 5 亿棵，这意味着光是这一个州便生产了全球近一半的咖啡。

20 世纪 30 年代，巴西咖啡农场的一名意大利劳工。

越来越多的土地被开垦，咖啡产量因而急剧增长。巴西的咖啡疆域越过圣保罗，向其南面和西面扩展，穿过中央高地来到其腹地，借助圣保罗广阔的铁路系统——其中就有咖啡专运路线——咖啡贸易被转移到了桑托斯港。

1905 年的农业普查数据充分抓住了圣保罗咖啡经济的特点。在该州 2.1 万座咖啡农场中，65% 的劳动力都来自外国。收入在前 20% 的农场主控制了这里 83% 的土地，生产了 75% 的咖啡，并雇用了 67% 的农业劳动力。最大的生产商是来自德国的弗朗西斯科·施密特，他拥有 700 万棵咖啡树，雇用了超过 4000 名工人。

不过，保利斯塔的农业体系并不是单一地种植咖啡。契约劳工移民普遍在种咖啡树的土地上种植粮食作物，许多大种植园经营的也都是混合农业。圣保罗州在粮食方面完全自给自足。[①]

稳定物价

1906 年，巴西在世界咖啡市场的优势地位达到了顶峰，咖啡产量为 2020 万袋，约占世界总产量的 85%。那年的大丰收迫使巴西对咖啡战略作出调整。在 19 世纪，巴西曾通过扩大生产，同时保持低价批发销售来刺激需求，从而提高咖啡收入。然而在世纪之交的彼时，供给超过了需求，咖啡价格随即从每磅 13 美分暴跌至 6 美分。

同年，圣保罗州政府出资专门购买过剩的咖啡，避免它们流入市场，以补贴身为德裔美国人的咖啡商赫尔曼·西尔肯领导的财团，该财团由银行家与经纪人组成。到了 1910 年，咖啡价格便恢复到每磅 10 美分以上，而到了 1913 年年底，西尔肯垄断集团手里的大部分咖啡都被卖掉了。

[①] Francisco Vidal Luna, Herbert S. Klein and William Summerhill, 'The Characteristics of Coffee Production and Agriculture in the State of Sao Paolo in 1905', *Agricultural History*, XC/1 (2016), pp. 22–50.

巴西当局这次精心安排的给咖啡"稳定物价"称得上咖啡史上的一个重大时刻：这是原产国首次对消费国的贸易条件进行控制。这在美国引起了公愤，西尔肯1912年因此遭到了国会委员会的审判。他解释道，如果没有巴西政府的这个计划，圣保罗就会爆发革命，但得到的回应毫无同情心可言："你认为那会比我们（美国）每磅咖啡要支付14美分的情况更糟糕吗？"[1]

此后，巴西当局经常通过"稳定物价"的手段来调节世界市场上的咖啡数量，将自身的出口价维持在每磅20美分以上。圣保罗州专门成立了一个机构来管理与咖啡相关的利益，这便是巴西咖啡管理局的前身。

20世纪30年代的经济大萧条打断了这一进程。当时面临的主要问题是，由于不断有新的土地投入生产，咖啡供应大幅增长，自1927年起，巴西每隔一年便有一次大丰收的记录。在那几年，仅巴西的咖啡收成就完全超过了世界市场的需求。到了1930年，巴西咖啡价格暴跌至每磅10美分以下。

巴西当局的反应可谓孤注一掷，而巴西咖啡管理局要设法应对这种局面。1931至1939年，巴西建立的75个大焚化厂构成了一个网络，将8000万袋咖啡（全球三年的供应量）化为乌有。对于新种植的咖啡树，该机构会进行相应的税务罚款；此外，它还为咖啡找到了别的用途，其中包

---

① Prendergast, *Uncommon Grounds*, p. 84.

括制作咖啡砖，为火车提供燃料。为促进消费，巴西咖啡管理局也试着打广告，以及在欧洲、俄罗斯和日本开设巴西咖啡馆。

为了缓解咖啡过剩的压力，人们会往火车发动机的蒸汽锅炉里倒咖啡与焦油的混合物，巴西，1932 年。

中美洲

造成巴西咖啡过剩的另一个原因是，哥伦比亚以及中美洲国家——哥斯达黎加、萨尔瓦多、危地马拉、洪都拉斯、尼加拉瓜、巴拿马和墨西哥——兴起，削弱了巴西在世界咖啡供应的主导地位。直到 1914 年，巴西咖啡的供应量还占美国咖啡进口量的 75%；而在两次世界大战期间，这一比例降到了 50% 左右。此外，来自其他国家的咖啡在美国咖啡交易所的价格明显溢价。

之所以溢价，是因为这些国家的咖啡品质优良，这意味着它们在咖啡培育与采收方面做得更加细致，采取的是湿处理法。这种处理要在水洗站或处理厂进行，处理厂是联系种植者与市场的中转站。大种植园往往自己经营加工厂，但小生产者则通常会直接向处理厂出售自家的咖啡樱桃。处理厂经常向种植者提供信用贷款，从而将他们与供应商交易有效地捆绑在一起。如此一来，处理厂的经营者便处于交易的优势地位，可以保证自己收到的都是优质成熟的咖啡樱桃，但这需要采摘者在整个采收季选择性地摘果。

中美洲国家离不开贸易出口带来的收入，因此鼓励生产者将咖啡树种植的范围扩大到偏远的、未开垦的高原地区，尽管那里往往有人居住。转变为商业咖啡种植需要整合土地，将其分配给私人，建立起规模足够庞大的劳动力队伍，这样才能生产出品质达标的咖啡。咖啡树大都由农民全年培育，无论是作为独立的所有者，还是签订各种租约，他们都是小规模种植，普遍以家庭为单位。[1]但问题是，从何处雇用采摘者。

各国根据自身情况采取不同的应对方式。在萨尔瓦多，关于流浪罪的相关法律迫使土著居民离开自己的土地，沦

---

[1] William Roseberry, 'Introduction', in *Coffee, Society and Power in Latin America*, ed. W. Roseberry, L. Gudmondson and M. Samper Kutschbach (Baltimore, MD, 1995), p. 30.

在哥斯达黎加，这种传统的彩绘牛车是咖啡种植者最初的运输工具，整趟旅程为期15天，咖啡从中央山谷的高原地区运到太平洋海岸的彭塔雷纳斯港，再从港口运往旧金山。2008年，彩绘牛车被列入了联合国教科文组织的《人类非物质文化遗产代表作名录》。

为种植园的劳动力。咖啡寡头阶层由此产生，他们有效地控制这个国家，结果整个20世纪萨尔瓦多不断出现各种不平等现象及冲突。1932年，贫苦的咖啡工人发起叛乱，导致政府军队对数万名萨尔瓦多土著居民展开"大屠杀"。

相反，在哥斯达黎加，政府通过了宅地法，允许定居者在几乎没有土著居民居住的高原地区对闲置的土地申请所有权。在咖啡处理厂的支持下，这些定居者建立了独立的小农场，而处理厂反过来又主要依靠来自伦敦的进口公司提供的信贷经营，它们所扮演的角色相当于哥斯达黎加咖啡的转口港。

危地马拉是首个在全球市场上给人留下深刻印象的中美洲国家，到 19 世纪末已成为世界第四大咖啡出口国。19 世纪 70 年代，在自由党领袖巴里奥斯将军的统治下，外国投资者可以在危地马拉购买规模庞大的种植园，同时也因为他们受到了《世界报》等欧洲报纸广告的诱惑。咖啡种植者钻法律的空子，让地方统治者强迫村庄提供劳动力，为自己争取季节性的采收工人。德国人被吸引过来，到了 20 世纪初，他们已拥有这里 10% 的咖啡农场，处理采收的 40% 的咖啡生豆，并控制了该国 80% 的咖啡出口。

第一次世界大战的爆发使欧洲咖啡市场受到严重影响。因此，中美洲国家进而重新定位，转向美国出口。其开始的标志是，旧金山咖啡经纪人克拉伦斯·比克福德与买家在 20 世纪一同进行咖啡样品杯测。如此便说明，仅凭颜色和大小对咖啡豆进行分类（如纽约交易所的做法）不足以确定其品质。小咖啡豆，如危地马拉的咖啡豆，原本折价交易——现在则享有溢价。[1]

旧金山的港口与成熟的铁路网络使旧金山成为中美洲咖啡在整个美国的分销中心。1914 年，巴拿马运河的开通加强了中美洲和拉丁美洲的太平洋咖啡出口港与北美洲和欧洲市场的输入港之间的联系。1913 年，美国从中美洲进口的咖啡总量为 3630 万磅；到 1918 年，便达到了 1.953 亿磅。

---

[1] Ukers, *All About Coffee*, p. 424.

哥伦比亚

"一战"后，哥伦比亚成为世界第二大咖啡生产国——产量从 1913 年的 6.1 万吨增至 1919 年的 10.1 万吨和 1938 年的 25.6 万吨。

据说，咖啡是由耶稣会牧师引入哥伦比亚的，他们中有人要求教区居民种植咖啡树以示忏悔。哥伦比亚咖啡种植区主要分布在安第斯山脉从北到南三条山脉的山坡上。复杂的地形对于铁路建设而言并不合算，因此产自这里的咖啡会由骡车队运到马格达莱纳河和考卡河，再被运往加勒比海的巴兰基利亚港和卡塔赫纳港，或者由空中缆车送至太平洋沿岸的布埃纳文图拉港。

19 世纪末，咖啡树的商业种植规模扩大。波哥大和麦德林的商人在桑坦德、昆迪纳马卡和安蒂奥基亚三省投资了咖啡种植园。受危地马拉成功经验的启发，他们采取了类似的种植技巧，即种植阴生植物以避免土壤侵蚀，有选择性地采摘成熟的咖啡樱桃并进行湿处理。他们主要以家庭为单位种植，实行不同的土地占有制——在桑坦德是收益分成制，安蒂奥基亚是佃农制，昆迪纳马卡则是众所周知的大庄园制。

20 世纪 20 年代，哥伦比亚的咖啡产量翻了一番，并且由于巴西"稳定物价"的手段而维持高价，因此哥伦比亚优质咖啡整体溢价超过 20%。咖啡树种植的边界扩展，南移到了卡尔达斯、托利马和乌伊拉三省。咖啡的出口量占该国总出口的 60% 至 80%，但在经济大萧条时期，哥伦比亚的咖啡产业深受巴西大丰收与随后的价格暴跌影响。由于土地所有者试图通过篡改与种植者的合同条款来转嫁自身损失，社会冲突随之爆发。许多纠纷逐渐演变成暴力，特别是在大庄园制的情况下。

关键时刻，哥伦比亚政府开始介入，在 1927 年成立了哥伦比亚国家咖啡生产者协会，使之扮演"私人实体的角色，为国家利益履行基本的公共职能"[1]。该协会的资金来自对每袋从哥伦比亚出口的咖啡所征收的税款，其职责是管理哥伦比亚的咖啡政策，维护种植者的"最大利益"。除了向其成员提供教育、金融和技术服务，该协会还管理哥伦比亚的咖啡出口贸易，并在国外推广哥伦比亚咖啡。

哥伦比亚，20 世纪 70 至 80 年代。人们在山坡上手工采摘咖啡樱桃。如图所示，对于哥伦比亚咖啡种植者而言，机械化并不在他们的选择范围内。

[1] Marco Palacios, *Coffee in Colombia*, 1850–1970 (Cambridge, 1980), p. 217.

哥伦比亚国家咖啡生产者协会拥有广泛权力，能够有效控制哥伦比亚咖啡与巴西咖啡之间的价格溢价。20世纪30年代，为了在美国获得更大的市场份额，协会曾故意缩小差价。到了1937年，哥伦比亚咖啡已占据美国市场的四分之一。

《美洲咖啡协议》

经济大萧条时期的咖啡价格暴跌，迫使拉丁美洲的主要咖啡生产商开始相互谈判，寻找解决危机的办法。1936年，他们成立了泛美咖啡局，促进了咖啡在美国的消费，而巴西咖啡管理局与哥伦比亚国家咖啡生产者协会之间签订了维持咖啡价格的协议，但随即瓦解。巴西指责哥伦比亚以扣压咖啡市场库存的方式"搭便车"，利用了巴西为调节咖啡供应所采取的措施。失望之下，巴西在1938年向市场投放了大量咖啡，从而令咖啡价格回落，但随着1939年欧洲开始爆发战争，各生产商急需找到可以避免价格进一步暴跌的办法。

1940年11月28日，西半球所有14个咖啡生产国与美国一同签署了《美洲咖啡协议》，这意味着这些国家认识到了保证咖啡稳定供应的重要性。该协议指出："有必要采取措施促进咖啡正常销售，这也相当合理，即根据需求调整供应，确保生产者和消费者享有公平的贸易条件。"①

各国代表咖啡生产者的机构就自身对美国的出口额展开谈判，该协议于1941年4月正式生效。到该年年底，咖啡价格便已翻了一番，并且此后一直保持较高的价格。

产生新的消费者

20世纪上半叶，美国的消费水平一路稳步提高。以1915至1920年为基准，到1946至1950年，美国咖啡的年均进口量翻了一番。即使是在经济大萧条时期，其发展也未中断。到了1939年，咖啡已成为一种家庭日常饮品——据报道，98%的美国家庭都会饮用咖啡。

美国社会的进一步发展为咖啡带来了新的机遇。1920至1933年的禁酒令促使咖啡馆取代了酒馆，因此咖啡——而非酒精——成为人们在外社交的合法饮品。而人们越来越强调在工作日吃所谓的简便午餐，结果在白天消费了更多咖啡。即便如此，家庭依旧是咖啡消费的主要场所。

但消费者想要的究竟是什么？

1924年，智威汤逊广告公司的研究调查表明，87%的家庭主妇认为风味是她们选择拼配咖啡最重要的因素。然

---

① Paul C. Daniels, 'The Inter-American Coffee Agreement', *Law and Contemporary Problems*, 8 (1941), p. 720.

而，"一般人很难明确区分咖啡风味有何不同"①。

得克萨斯州零售商哈里·朗格表示，咖啡市场主要面向四类消费者——都是家庭主妇——并打出"任意拼配"的旗号来吸引各类消费群体。②

"对咖啡无所不知的主妇"，奈何找不到

适合自己口味的咖啡：

咖啡风味提高，饮食愉悦感增强

桌角的咖啡壶是平衡晚餐水平的关键。

如果咖啡不知何故"有点发挥失常"——可能是咖啡自身的问题——整顿饭吃起来就不尽如人意；但当它"符合口味"时，这顿饭便自始至终都令人愉悦。

如若对找到"平衡点"感到困扰，

请选择"任意拼配"咖啡，

由它为您来作适当调整。

初为人妻的主妇，对咖啡了解甚少，

但想找到自己与丈夫

可以信赖的优质拼配咖啡：

一次正确的选择

首次购买便选择"任意拼配"咖啡，

每天清晨都可以来一杯。不少家庭长时间

① Prendergast, *Uncommon Grounds*, p. 157.
② Ukers, *All About Coffee*, p. 484.

都选不对咖啡，当然，那是因为他们还不了解

"任意拼配"咖啡——

即便现在，除非亲自尝尝，否则很难真正了解它。

这就是我们会一再希望

您先订购一磅品尝一下的原因。

对自己购买的咖啡感到满意的主妇：
省心的服务

想买咖啡时，选择"任意拼配"咖啡。

买"任意拼配"咖啡，够省事！

省心：风味与烘焙程度保持不变。

省时：订购"任意拼配"咖啡，咖啡豆的研磨程度

可以根据顾客自己的渗滤壶或咖啡壶的规格要求。

省钱：清楚知道每次制作若干杯咖啡所需的量，

可以避免浪费。

对于家有雇工的主妇：
你能说得出家里的咖啡的名称吗？

还是说，那是照厨师的意见，从店里买来的

不知名品牌的咖啡？

选择让厨师采购，若对方值得信赖，便算走运，

但最好告诉对方，比起现在用"不知名品牌"

制作的拼配咖啡，

自己更喜欢"任意拼配"咖啡。

后者的一大优点便是：刚刚烘焙好。

告诉厨房女佣，现在就订购"任意拼配"咖啡吧。

朗格的这些话术都是在利用消费者对咖啡缺乏信心的心态。在外人看来，咖啡就代表着家庭，因此他实际制造了人们对于咖啡品质的焦虑。选择对的咖啡被认为对家庭和谐而言至关重要——没有哪个新婚妻子愿意生活在一个"选择错误的家"。

品牌与广告

到 20 世纪 30 年代末，90% 以上的烘焙咖啡在出售前都已称好重量，并且包装上带有商标。全球有超过 5000 个咖啡品牌，但领先的三家——A&P、麦斯威尔和蔡斯与桑伯恩——占据咖啡市场 40% 的份额。这些品牌之所以占据主导地位，部分是因为一半以上的购买量都出自杂货连锁店，如 A&P 旗下的杂货店。到了 1929 年，麦斯威尔和蔡斯与桑伯恩两个咖啡品牌分别被通用食品公司和标准品牌公司收购——这两大巨头利用自身经济实力，确保这两个品牌的咖啡能够摆在超市货架上的显眼位置。

咖啡生产商利用两次世界大战间隔年间开发的大众渠道，展开了强有力的传播活动。广告公司开展各种宣传活动，例如麦斯威尔咖啡在时尚杂志上刊登的广告，重点展现了特色精致的咖啡馆以及西奥多·罗斯福总统对麦斯威尔咖啡所谓的代言——"滴滴香浓，意犹未尽"。1933年，麦斯威尔赞助的广播综艺节目《麦斯威尔演艺船》一经推出，很快便成为美国最受欢迎的节目之一。节目中，好莱坞各大明星通常是一边喝着咖啡，一边与节目主持人聊天，中间还穿插着音乐与表演，同时提醒听众"只需选择购买麦斯威尔咖啡，便可收听节目"。在节目推出后的一年里，麦斯威尔咖啡的销售额增长了85%。[1]

大烘焙商的很多宣传信息都是在利用前面朗格发现的人们对咖啡的不安感。蔡斯与桑伯恩咖啡经常宣传的广告内容是妻子因为没有提供令人满意的咖啡而被丈夫责备。诸如此类的广告意在教育读者，鼓动他们购买"带有日期的"咖啡（即印有商店到货日期）并购买真空包装的咖啡，这种咖啡的新鲜度有保障。即使如此，作为这种包装方式开创者的希尔斯兄弟公司也发表过免责声明："只要咖啡交付时完好无损，我们的责任便到此为止，接下来只有您以恰当制作咖啡的方式与我们合作，我们的付出和您的金钱才不会浪费。"[2]

"二战"期间及战后

美国加入第二次世界大战后，曾在1942至1943年间有

---

[1] Prendergast, *Uncommon Grounds*, pp. 193–6.
[2] Steve Lanford and Robert Mills, *Hills Bros. Coffee Can Chronology Field Guide* (Fairbanks, AK, 2006), pp. 19–25.

过短暂的咖啡配给期，而这场战争也促进了咖啡的普及。

当军官意识到提高士气的重要性后，便会鼓励手下士兵饮用咖啡，因此士兵成了狂热的消费者。和在内战时期一样，咖啡似乎起着刺激和安慰的作用，或许还能调节单调无味的生活。"二战"后，一项针对海军的早期战后研究表明，水手在航海过程中的咖啡消耗量是普通百姓的两倍——即便是陆上海员的消耗量也比全国平均水平高出50%。[①]

事实证明，新增"咖啡时间"可以使军需人员提高工作效率。因此，"咖啡时间"在全军推广开来。而在战后，这一安排还扩展到了普通百姓的生活中，到了 20 世纪 50 年代中期，大约 60% 的美国工厂都设有"咖啡时间"。这在一定程度上归功于泛美咖啡局对在工作场所安排"咖啡时间"的大力支持与推广。它还提倡"在路上的咖啡时间"，认为随着美国机动车辆增多，咖啡能让司机保持警惕。

1954 年冬季的一项调查结果表明，消费者平均每天会饮用两杯半咖啡。其中两杯在家饮用——通常是在早餐和晚餐时间——其余则是在咖啡馆 / 餐厅或工作时饮用。城镇居民每天饮用 2.8 杯咖啡，农村居民的饮用量平均为 2.3杯。然而，咖啡消费水平最高的地区是中西部的农业区，这或许也反映了这里许多居民都有斯堪的纳维亚血统。

---

① Andrés Uribe, *Brown Gold: The Amazing Story of* Coffee (New York, 1954), pp. 42–4.

一个时代的终结

第二次世界大战刚结束时，美国 10 岁以上人群的人均咖啡饮用量超过 8.6 千克（约合 19 磅），达到了历史顶峰。拉丁美洲的咖啡产量占世界产量的 85%，其中 70% 都出口至美国，事实上，如今几乎每个美国家庭都饮用咖啡。美国"一杯咖啡"[1]的概念——该词在 20 世纪 30 年代首次出现，就是"普通咖啡"的意思——已牢牢树立起来。它指的是一种醇厚度较低、风味较淡的咖啡，以相对较大的量作为佐餐饮品。其口感不仅反映了巴西咖啡豆的平淡无味——

1941 年，美国一家餐厅的女服务员正在为顾客倒咖啡。美国经典的"一杯咖啡"是用手冲咖啡设备制作的，然后在加热板上加热，可以无限续杯。

---

① 原文为"cup of Joe"，为"一杯咖啡"的俚语，Joe（乔）是普通人名。——译注

由于用渗滤壶冲泡，导致咖啡过度萃取——也反映了美国家庭主妇对咖啡用量的吝惜。

然而到了 20 世纪 50 年代末，随着年青一代对软饮料的喜爱，美国咖啡的消费水平已明显下降，欧洲即将超过北美，成为咖啡最大的消费市场。与此同时，拉丁美洲的咖啡生产商再次面临供过于求所造成的低价危机，而非洲和亚洲的新兴生产商选择种植价格更为便宜的罗布斯塔咖啡则使情况进一步恶化。咖啡已经成为一种全球商品。

［第五章］

全球商品

20 世纪下半叶，咖啡成为全球商品。出现这一现象的根本原因在于人们改种适应能力更强的罗布斯塔咖啡树来替代阿拉比卡咖啡树，非洲和亚洲的咖啡生产得以振兴。罗布斯塔咖啡更廉价，可以带动新的消费者日常饮用咖啡，也极大地改变了咖啡的口味与形式。国际制度的完善促进了世界咖啡市场秩序的规范化，但事实证明，它依旧无法使咖啡生产者免受价格波动的影响，最终在 20 世纪末爆发了世界咖啡危机。

罗布斯塔咖啡与非洲复兴

20 世纪初，荷属东印度群岛从比属刚果引入罗布斯塔咖啡树，取代因咖啡叶锈病而受损的阿拉比卡咖啡树。到了 20 世纪 30 年代，东印度群岛 90% 以上的咖啡都是罗布斯塔咖啡。该品种的咖啡在美国烘焙商中十分畅销，因为他们可以借此宣传自家的拼配咖啡中有来自爪哇岛或苏门答腊岛的咖啡。第二次世界大战以及之后一系列的独立战争导致印度尼西亚的咖啡生产遭受重创。直到 20 世纪 80 年代，印度尼西亚才再次成为世界最大的罗布斯塔咖啡生产国。

1960—2019 年主要咖啡生产国（按年代划分）[1]

1960—1969	1970—1979	1980—1989	1990—1999	2000—2009	2010—2019
巴西	巴西	巴西	巴西	巴西	巴西
哥伦比亚	哥伦比亚	哥伦比亚	哥伦比亚	越南	越南
安哥拉	科特迪瓦	印度尼西亚	印度尼西亚	哥伦比亚	哥伦比亚
乌干达	墨西哥	墨西哥	越南	埃塞俄比亚	印度尼西亚
科特迪瓦	印度尼西亚	科特迪瓦	危地马拉	印度	埃塞俄比亚
墨西哥	埃塞俄比亚	埃塞俄比亚	印度	墨西哥	印度

非洲自摩卡咖啡衰落后便无人问津，而罗布斯塔咖啡使之重新占据全球咖啡经济的中心地位。非洲大陆 1914 年的咖啡产量只占世界产量的 2%，而相比之下，1965 年的占比是 23%。[2]其中 75% 的产量都来自罗布斯塔咖啡，罗布斯塔咖啡树主要种植于法国和比利时在西非与中非的前殖民地，以及乌干达和安哥拉两国。

科特迪瓦的咖啡产量在 1939 年还不足 1.6 万吨，而到了 1958 年逐渐增至 11.4 万吨。1960 年独立后，其咖啡产量迅速增长，在 1970 年便达到了 27.95 万吨。在接下来的 20 年里，产量水平保持稳定。20 世纪 70 年代，科特迪瓦成为世界第三大咖啡生产国（仅次于巴西和哥伦比亚），也是罗布斯塔咖啡的主要出口国。而这都是该国第一任总统费利

---

① Data courtesy ICO, 1965/6 to 2016–17. Note 1960s = crop years 1965– 6 to 1969–70; 2010s = crop years 2010/11 to 2016/17.

② Stuart McCook, 'The Ecology of Taste', in *Coffee: A Comprehensive Guide*, ed. R. Thurston, J. Morris and S. Steiman (Lanham, MD, 2013), p. 253.

1924 年，在苏门答腊岛一座种植园的橡胶树荫下的罗布斯塔咖啡树，该种植园为荷兰人所有。

科特迪瓦通过自然处理法干燥的咖啡樱桃

克斯·乌弗埃–博瓦尼的功劳。

乌弗埃–博瓦尼曾是位种植咖啡树的农民，他发起运动，反对法国种植园主在殖民时代的特权，特别是他们对非自愿劳动力的剥削。这些特权一被废除，科特迪瓦当地种植者的生产效率更高了。在科特迪瓦独立后，乌弗埃–博瓦尼鼓励同胞们不要"无所事事、故步自封"，而要专注于种植品质优良的咖啡树，实现"发家致富"。[①]他继续沿用了法国殖民者曾设立的农产品价格稳定和支持银行。该机构会制定咖啡在各个阶段的购买与销售价格。它将咖啡的生产商、加工商以及出口商联系起来。虽然该国的咖啡贸易仍掌握在私人手中，但科特迪瓦政府通过制定保护价格来维护咖啡种植者的利益，从而有助于在中部林区开拓新的咖啡产地。

非洲几乎所有讲法语的国家都设立了农产品价格稳定和支持银行。支持该机构运作的资金来自出口商利润的所得税。它旨在在全球价格偏高时积累收入，以在价格偏低时维护生产商的利益，并为提高生产力和促进农业多样化提供资助。该机构在科特迪瓦的运作相对较好。在1974至1982年间，农户种植咖啡的出口价格差不多是世界价格的70%。但对许多国家而言，压低支付给国内生产商的价格，用产生的盈余来资助非营利性政府项目，这种诱惑显然太大了，因此农产品价格稳定和支持银行经常沦为制度腐败

---

① Jennifer A. Widner, 'The Origins of Agricultural Policy in Cote d'Ivoire', *Journal of Development Studies*, XXIX/4 (1993), pp. 25–59.

之源。

从前被英国殖民统治的非洲国家保留了由英国殖民当局引入的营销委员会。这些机构先购买该国处理好的咖啡豆以供出口，然后将自己收到的交换物返给该国。营销委员会要负责咖啡的分类、分级和拼配。通过差别定价，即对可列为优质批次的咖啡提高定价，从而鼓励优质咖啡的生产。

20世纪初，白人移居者——其中最著名的便是是卡伦·布利克森——在肯尼亚建立农场。他们在此种植阿拉比卡咖啡树，但采用的并非源自埃塞俄比亚的铁皮卡，而是波旁品种。这部分解释了埃塞俄比亚咖啡与肯尼亚咖啡在风味上的差异，前者具有花香与柑橘气味，而后者则是葡萄果酱与黑莓味道。20世纪50年代茅茅起义①爆发之后，肯尼亚开始推行农业改革，鼓励建立家庭农场，将发展自给农业与种植经济作物——特别是咖啡——相结合。1963年肯尼亚独立后，政府保留了咖啡委员会的核心拍卖制度，因此出口商可以根据咖啡杯测结果，选择购买不同等级的咖啡，种植者则按该等级咖啡的平均价格出售，有利于促进优质咖啡的生产。相比之下，在新独立的坦桑尼亚，咖啡由咖啡委员会以相同批次出售，很快便声誉扫地，直到20世纪90年代中期改革后才恢复。

与此同时，乌干达也成为罗布斯塔咖啡的主要生产国，

---

① 茅茅起义（Mau Mau uprising），指1952至1960年肯尼亚人民反对英国殖民者的武装斗争运动。——译注

大约在20世纪80年代，肯尼亚对咖啡生豆样品进行分级。1963年肯尼亚独立后，咖啡之所以能够成为该国外汇收入的主要来源之一，离不开对咖啡质量的严格把控。直到2002年，肯尼亚人才能购买和消费自己国家种植和烘焙的咖啡。

其产量从 20 世纪 40 年代末的 3.1 万吨增至 1962 年[1]独立时的 11.9 万吨。到了 1969 年，乌干达的咖啡产量更是达到了 24.7 万吨的峰值，其中绝大部分都是由诸如维多利亚月牙湖等地区的小农户在园地种植的。与众不同的是，乌干达的生产商对罗布斯塔咖啡豆进行水洗处理，提高了咖啡品质。由于伊迪·阿明独裁统治下实施的糟糕政策以及之后十年政治与军事动乱，乌干达的咖啡产量从 20 世纪 70 年代中期开始下降。尽管咖啡收入占乌干达出口收入的 90% 以上，但该国的咖啡营销委员会已逐渐沦为一个臃肿的官僚机构，给种植者的出口价格甚至低于咖啡市场价格的 20%。[2]

速溶咖啡

1929 年，巴西当局急需为本国产量过剩的咖啡豆寻找其他用途，于是前去询问瑞士跨国食品制造企业雀巢公司是否可以为其开发一种浓缩咖啡固体饮料。雀巢公司从事研究的科学家马克斯·莫根塔勒花了六年多的时间，开发出了一种美味可口的速溶咖啡，不过当时巴西对此早已不感兴趣，他自己的研究团队也都已撤走。[3]

雀巢咖啡在 1938 年推出，是一种通过喷雾干燥技术而

---

[1] 此处原文有误，经查证，应为 1963 年。——译注

[2] Moses Masiga and Alice Ruhweza, 'Commodity Revenue Management: Coffee and Cotton in Uganda', *International Institute for Sustainable Development* (2007).

[3] Nestlé, *Over a Cup of Coffee* (Vevey, 2013), pp. 25–30.

获取的咖啡提取物。后来欧洲爆发战争，导致其销售主要集中在美国，美国陆军部由于军需几乎购买了市面上的所有咖啡。战争结束后，通过美国士兵的背包和美国援助欧洲合作组织派送的包裹，咖啡又重返欧洲。1965 年，采用冷冻干燥技术的罐装优质产品"雀巢金牌"咖啡问世。

与此同时，美国主要的烘焙商也开始生产速溶咖啡。1953 年，麦斯威尔速溶咖啡在美国的产量超过了雀巢咖啡。到了 20 世纪 50 年代末，速溶咖啡已占美国咖啡市场份额的 20%，主要集中于食品杂货业的低端市场。和雀巢咖啡一样，麦斯威尔速溶咖啡一半以上都采用了罗布斯塔咖啡豆的拼配咖啡。

接下来，咖啡市场欠发达的国家开始以速溶咖啡来塑造国内的咖啡口味偏好和消费习惯。在饮茶为主的英国，由于恰逢商业电视的出现，咖啡消费量在 20 世纪 50 年代翻了一番。因为观众发现，电视广告的时长恰好足以让他们冲杯速溶咖啡，但来不及按传统方式泡杯茶。到了 20 世纪 90 年代，咖啡在英国的价值（而非销量）超过了茶，其中速溶咖啡占销售总额的 90％。1987 年，雀巢咖啡推出的经典的"雀巢金牌咖啡情侣"系列电视广告——以两位互为邻居的男女主角之间朦胧暧昧的关系为主题——巩固了其在英国市场的主导地位。1992 年 12 月，半数以上的英国人收看了该系列的最后一集，据说这让雀巢咖啡的销售额

"啊——闻着就像新鲜烘焙的全豆咖啡。"雀巢咖啡的这则广告展示了最早的玻璃罐装产品之一。

增加了 70%。[1]

1957 年，雀巢公司的代表在希腊塞萨洛尼基的商品交易会上，将速溶咖啡粉与冷水在可可饮料摇瓶中混合，由此产生了一层厚厚的泡沫。然后用更多的水稀释，再加冰，最后喝起来的口感非常清爽。于是该公司开始推广这种新的咖啡制作方式，受到了希腊年轻人的追捧，成了户外生活方式的象征。如今，这种名为法拉沛的冰咖啡成为风靡全希腊的夏季饮品。[2]

欧洲咖啡风尚的兴起

罗布斯塔咖啡的出现明显改变了"二战"后欧洲其他国家的咖啡口味偏好。到了 1960 年，罗布斯塔咖啡已占法国咖啡消费量的四分之三，因此为了中和这种咖啡豆的苦味，必须将之深度烘焙到出现焦糖化反应。即便在荷兰与比利时，最受欢迎的烘焙方式也是中度烘焙，这反映了它们与罗布斯塔咖啡产地早先的殖民关系。意大利的咖啡偏好则呈地区性分布，越往南越喜欢罗布斯塔咖啡。虽然一开始，这体现的是这种咖啡的廉价性，但它所形成的消费者偏好不容冒犯。葡萄牙推出了本国浓缩式咖啡饮品——比卡，采用产自安哥拉的罗布斯塔咖啡拼配冲泡，该产地

---

[1] Claire Beal, 'Should the Gold Blend Couple Get Back Together?', www.independent.co.uk, 28 April 2010.

[2] Vivian Constantinopoulos and Daniel Young, *Frappé Nation* (Potamos, 2006).

的咖啡生产主要来自白人经营的种植园，它在20世纪70年代初的产量达22.5万吨，直到安哥拉后来爆发了一系列独立战争，中断了整个咖啡业的运作。

20世纪，随着欧洲咖啡业的发展，市场采用各种形式的咖啡制作技术、烘焙程度、拼配方式以及消费习惯，从而形成了特色鲜明的"各国口味"。如今，许多与某些特定国家相关的"传统"咖啡风尚都可追溯至20世纪。一般而言，这些风尚只会在"二战"后形成的大众消费社会中不断发展。随着小杂货店与精品咖啡馆的衰落，地方及区域烘焙商逐渐被超市取代。超市出售的咖啡品牌极具辨识度，购物者可以从花费高昂的电视广告中一眼认出，这些广告宣传的都是"各国"主要的烘焙商。各阶层的消费方式类似，喝咖啡成了代表"各国身份"的日常象征。

德　国

德国在1871年实现统一，此后便成为欧洲最大的咖啡市场。到了20世纪，德国所有地区、各个阶层的人们日常饮用咖啡，它是家庭成员早晚享用面包或土豆时的温暖佐餐饮品。德国对咖啡的庞大需求，其中很大一部分都要由咖啡替代品来满足：1914年，德国在消费1.8亿千克（约合4亿磅）咖啡的同时，也消费了1.6亿千克（约合3.5亿磅）

的咖啡替代品。[1]

各种喝咖啡的习惯随之兴起，例如咖啡午后派对——女性午后一边闲聊一边享用咖啡与蛋糕的聚会。德国各大城市的咖啡摊生意兴隆，人们在前往工厂或办公室上班途中，或者休息时会在此购买咖啡。在周日的家庭聚会上，饮用的都是"优质"咖啡——用真正的咖啡豆，而非咖啡替代品制作而成；夏日午后，最美妙的事情莫过于到当地公园里的咖啡园散散步。其中一些还有热水出售，供不太富裕的顾客冲泡自己带来的咖啡。

汉堡成为欧洲主要的咖啡港口，如今这里进口的咖啡大约90%来自拉丁美洲，其中许多都是通过德国移民企业建立的运输网航运。[2]20世纪初，于1887年成立的咖啡交易所已有大约200个由进口商、经纪人和批发商组成的成员组织。仓库女工会对托运的货物分级挑选，并将发货样品置于港口专门放咖啡的"邮箱"内。铁路网络可将咖啡运送给德国各地的众多批发商和杂货店。

19世纪80年代，位于莱茵兰的范·居尔彭、伦辛与冯·金博恩机械工程公司，即后来的普罗巴特公司，开始生产滚转炉。生产这些设备是为了开创咖啡烘焙批发业务，当时主要向当地杂货店供应散装咖啡豆。烘焙商们直到后来才

————————

[1] Julia Rischbieter, '(Trans)National Consumer Cultures: Coffee as a Colonial Product in the German Kaiserreich', in *Hybrid Cultures–Nervous States*, ed. U. Lindner et al. (Amsterdam, 2010), pp. 109–10.

[2] Dorothee Wierling, 'Coffee Worlds', *German Historical Institute London Bulletin*, XXXCI/2 (November 2014), pp. 24–48.

建立自己的品牌身份，不过大多仅限于区域市场。

　　但也有例外，主要是那些经营规模较大的直接分销业务的品牌，如自有品牌食品连锁店恺撒咖啡，它在"一战"前夕便拥有 1420 家商店。20 世纪 30 年代，位于不来梅的埃杜斯科邮购运营商成为德国最大的烘焙商。1949 年成立

在 20 世纪 80 年代的西德，咖啡品鉴师正在为世界最大的烘焙商之一雅各布斯检查咖啡烘焙样品。

于汉堡的邮购运营商奇堡在"二战"结束后不久，便摇身一变成为西德最大的烘焙商，并在20世纪五六十年代发展了一系列的小型连锁店。人们可以在这些店里品尝咖啡，以及为家里购买咖啡豆。20世纪70年代，奇堡公司进行业务扩展，在面包店内和超市里面售卖咖啡。1997年，奇堡与埃杜斯科合并为一家公司。

过滤式咖啡越来越受德国人喜爱。1908年，德累斯顿家庭主妇梅利塔·本茨为一种新的咖啡制作方法申请了专利，该方法就是将滤纸放入圆孔黄铜滤壶中。据说，她是用儿子学校的吸墨纸做实验的。在这之前，咖啡过滤都要靠洗得干干净净的布，并且还得重复使用。如今，家庭主妇们只需扔掉滤纸与滤渣，便算清理完毕。梅利塔·本茨的丈夫成立了一家以她的名字命名的公司，很快便大获成功，并在20世纪30年代推出了如今为人熟知的锥形过滤器与滤纸，从而稳固了自身地位。

斯堪的纳维亚

同样，北欧国家也发展出了具有强烈地域色彩的咖啡文化。这是咖啡作为御寒饮品的功能优势和禁酒运动两相结合的结果。禁酒运动与教会密切相关，而教会将咖啡推广为酒饮的替代品。丹麦与瑞典的人均咖啡消费量在20世纪30年代便已超过美国。到了20世纪50年代，世界人均咖啡消费量最大的国家是芬兰，挪威次之。

20 世纪 50 年代，萨米族的驯鹿牧民每天通常会喝 12 杯咖啡。男人在早晨出门前会喝杯咖啡取暖。当听到宣告男人回家的狗吠声时，女人就开始现煮新鲜咖啡。牧民热情待客的礼仪总是围绕喝咖啡展开。这意味着每次有人拜访，主人会奉上，客人也会喝上至少两杯咖啡，这也可能使他们每天的咖啡饮用量增至 20 杯。[①]

城市社区也围绕咖啡展开了各种日常活动和仪式。19 世纪末，咖啡小休——与家人、朋友或同事分享咖啡和蛋糕的时间——在瑞典逐渐发展起来。它在瑞典文化中占据了核心地位，如今就连难民也开始参与其中。丹麦语中有专门的术语来描述为分娩妇女和护理助产士准备的咖啡。芬兰劳动法还正式规定了工作日内的咖啡时间。[②]

斯堪的纳维亚人较为喜欢轻度烘焙的咖啡。于是该行业的烘焙师延续了这种偏好，以反映其民族特性。20 世纪 20 年代，芬兰的主要烘焙商保利格在自家品牌的咖啡产品上印了一个身着民族服装、正从壶里倒咖啡的年轻女孩的图像。自 20 世纪 50 年代，保利格便开始挑选一位年轻女性作为"保拉女孩"，让她在公众面前推广自家品牌。

---

① L. Whitaker, 'Coffee Drinking and Visiting Ceremonial Among the Karesuando Lapps', *Svenska landsmål och svensktkt folkiv* (1970), pp. 36–40.

② Dannie Kjeldgaard and Jacob Ostberg, 'Coffee Grounds and the Global Cup: Global Consumer Culture in Scandinavia', *Consumption, Markets and Culture*, x/2 (2007), pp. 175–87.

50 多年来，用铜壶上提神咖啡的"真正的""保拉女孩"。上图为自 1962 到 1969 年担任该职的第三代"保拉"阿尼亚·穆斯塔迈基。

意大利

由于自身浓缩咖啡冲泡技术的发展，意大利逐渐形成了独特的欧洲咖啡文化。[①]高级鸡尾酒酒吧可以迅速备好饮品，然后通过柜台传给顾客，这种酒吧的扩张使餐饮业需要一种同样快速供应咖啡的方式。若是在冲泡过程中增加一定的压力，加快萃取时间，便能为每个顾客"专门"准备一杯新鲜的咖啡。世界第一台商用咖啡机拉帕沃尼"理

---

① Jonathan Morris, 'Making Italian Espresso, Making Espresso Italian', *Food and History*, VIII/2 (2010), pp. 155–83.

想"咖啡机诞生于 1905 年，在米兰制造。它由一个锅炉组成，用蒸汽带动热水向下冲，通过置于输送口（"炉顶"）的咖啡。不过由于这种机器的压力相对较小（1.5 至 2 帕），整个准备过程仍要大约一分钟的时间，才能做出浓缩的过滤式咖啡的味道。这些大型机器具有很强的装饰效果，往往被放置在欧洲许多高级酒店的咖啡吧柜台上。意大利法西斯政府怀疑咖啡属于"外国奢侈品"，因此普通意大利人更常接触的是咖啡替代品。

1948 年后，这种情况有所改变，阿希尔·加贾制造了一台全新的浓缩咖啡机，利用与弹簧活塞相连的杠杆将水冲入咖啡。这种机器达到的压力更高（约 9 帕），输送速度更快（约 25 秒），由此生成的浓缩液表面覆有一层咖啡脂泡沫，即克丽玛。随后，以飞马为代表的制造商推出了半自动咖啡机，以电动泵取代活塞。如今，咖啡吧的咖啡无论外观还是口味都不同于家里制作的。卡布奇诺也同样如此，它原本是指加奶咖啡，但现在专指加了蒸汽牛奶的浓缩咖啡，且仅在咖啡吧出售。

20 世纪 50 至 60 年代，现代意大利咖啡文化开始兴起。工业化和城市化使街区咖啡馆增多，它们服务于随农村迁往城市的移民而出现的小作坊与住宅区。由于准备和饮用咖啡的速度提高，咖啡馆成为人们工作前喝杯卡布奇诺及白天稍作休息的理想场所。站着喝咖啡变成大众饮用咖啡的标准做法，因为 1911 年议会出台了一项法规，规定了"一杯咖啡，不带服务"——站在咖啡柜台前饮用的咖啡——

的最高价格。为了抑制通货膨胀，咖啡价格通常定得不高，导致咖啡吧对于连锁企业而言毫无吸引力。

浓缩咖啡工艺的关键优势之一在于强化风味，因此价格较为便宜的商品咖啡豆可以成为拼配咖啡的重要组成部分。"二战"后，巴西向意大利出售了大量低品质的桑托斯咖啡豆，于是意大利烘焙商也开始使用罗布斯塔咖啡豆，其额外的好处是可以制作出更浓密、看上去诱人的克丽玛。

1955 至 1970 年，意大利国内咖啡消费量翻了一番。众所周知的摩卡壶，即比乐蒂公司制造的铝制八角炉灶咖啡机，已成为意大利厨房的标配。它的主要功能是渗滤：蒸汽压力迫使下腔室的热水上升，并通过咖啡粉，最后在顶部的供应区收集咖啡液。该咖啡机的广告是可以制作出"恰

意大利罗马，1957 年。咖啡吧侍者正在使用加贾咖啡机，这是全世界第一台高压咖啡机，可以在浓缩咖啡表面覆上一层咖啡脂。正如机器正面标语所示，这被称为克丽玛咖啡。

如咖啡吧里的咖啡"，即便没有制作出克丽玛。

20世纪60年代，来自皮埃蒙特的拉瓦萨公司成为意大利首家全国性的咖啡生产商。它的成功源自开创性的电视广告活动，使用卡通人物宣传，结合分布广泛且已渗透到全国各地多个街区商店的分销体系。1995年，自路易吉·拉瓦萨在都灵开设第一家烘焙咖啡杂货店一百年后，该公司在意大利"家庭"市场上占有45％的份额。

中　欧

20世纪初，维也纳咖啡馆的火爆程度达到了顶峰。咖啡馆的成功恰好伴随着文化与消费的民主化，而后者也正是欧洲在19世纪末的特点。到了1902年，维也纳城市内部大约有1100家咖啡馆，吸引着广大的中产阶级顾客，同时还有4000多家工人阶级光顾的酒馆。[1]

咖啡馆的公共开放性是它吸引维也纳犹太人等群体的关键因素，他们当时在奥地利社会仍面临各种偏见。吸纳了许多犹太作家的"青年维也纳"文学圈子通常在格林斯坦咖啡馆聚会。包括列夫·托洛茨基在内的社会主义思想家们经常光顾中央咖啡馆。这些小团体的聚会通常有固定桌位，即咖啡馆专门为那些每天随时会来的常客保留的桌位。侍者大多是男性，由一位被称为"服务员"的人主管，

---

[1]　Charlotte Ashby, Tag Gronberg and Simon Shaw-Miller, eds, *The Viennese Café and Fin-de-siècle Culture* (London, 2013).

这意味着咖啡馆基本不会被误认为是名声不好的地方。女性客人很受欢迎，但由于咖啡馆幽暗的室内环境及男性化的氛围，很多女性更愿意在糕点咖啡店喝咖啡。

维也纳咖啡馆兴起的现象后来遍及整个奥匈帝国。20世纪 30 年代初，布达佩斯约有 500 家营业的咖啡馆，其中最漂亮的莫过于 1894 年成立的纽约咖啡馆。作为帝国在亚得里亚海的出海口，的里雅斯特成为欧洲主要的咖啡港口之一，第一次世界大战后被划分给了意大利，但此后仍旧保持着咖啡大港的地位。而出生于匈牙利家庭（位于今天罗马尼亚的蒂米什）的费伦茨·伊利能在 1933 年创立意大利最大的咖啡烘焙公司之一，便是因为"一战"期间他曾在奥匈帝国军队服役，退役后就留在了的里雅斯特。

匈牙利布达佩斯，2006 年。修复后的纽约咖啡馆内部，首次开业时间是 1894 年。

成立于 1862 年的维也纳殖民商品店的老板尤利乌斯·迈因尔开设了一家烘焙公司,后来在儿子尤利乌斯·迈因尔二世的领导下,该公司发展成为中欧最大的咖啡供应商。到了 1928 年,它在奥地利、匈牙利、捷克斯洛伐克、南斯拉夫、波兰和罗马尼亚共经营着 353 家杂货店。1938 年,选择与犹太人结婚的著名反纳粹分子尤利乌斯·迈因尔三世举家迁往伦敦,直到战争结束后才回去,为的是让公司重整旗鼓。

奥地利咖啡馆设计了囊括丰富咖啡饮品的菜单。除了黑咖啡、棕色咖啡、黄金咖啡和牛奶咖啡(从名字便可看出牛奶与咖啡混合的相对比例),还有更小众的咖啡,其中包括马车夫咖啡(杯中的黑咖啡表面放了大量攒奶油)和斯佩贝尔-土耳其人咖啡(制作时加一块糖的双倍量土耳其咖啡,最早由一位著名的律师饮用)。[①]然而,从 20 世纪 50 年代开始,维也纳咖啡馆迅速采用了浓缩咖啡机,于是牛奶咖啡和卡布奇诺这两种咖啡变得十分相似。

在中欧其他地方,咖啡文化总是离不开共产主义,反之亦然。在捷克斯洛伐克等国出售的"标准混合咖啡",其来源及具体混合的东西往往令人心生疑虑:1977 年,东德为了应对货币危机,推出了一款混合咖啡——将咖啡粉与烤豌豆、黑麦、大麦和甜菜混合。在匈牙利,街区的浓缩咖啡吧的兴起意味着对已然根深蒂固的咖啡文化的让步,

① Harold B. Segel, *The Vienna Coffeehouse Wits*, 1890–1938 (West Lafayette, IN, 1993), p. 11.

但直到 1989 年后，纽约咖啡馆及其他在布达佩斯的咖啡馆才恢复昔日辉煌。柏林墙倒塌后，东欧和中欧的咖啡消费量开始大幅上涨。

日 本

20 世纪下半叶，欧洲和北美以外的消费市场在全球咖啡贸易中占有重要地位。作为主要推动力，便利品使长期形成的精英咖啡文化向更广泛的群体开放。

作为当今世界第三大进口国[①]的日本便是一个例证。17 世纪末，咖啡由荷兰东印度公司首次引入日本。但咖啡消费仅限于长崎附近的人工岛，因为在德川幕府时期，日本还在"闭关锁国"，只通过该岛进行对外贸易。这一时期，唯一喜欢喝咖啡的日本人便是岛上的妓女，她们珍视咖啡，是因为它能使之保持清醒，防止客人不付钱就离开。[②]

在"明治维新"运动以及美国要求日本开放国际贸易的压力下，咖啡在 19 世纪后期进入了日本社会。明治二十一年（1888 年），郑永庆[③]以自身在纽约和伦敦看到的咖啡馆为原型，创立了"可否茶馆"。该店仿照精英俱乐部的模式，设有皮扶手椅、地毯、台球桌和大量写字台，还放有各种报纸书刊。可惜，以一杯咖啡的价格换取使用这

---

① 此乃作者写作本书时的统计数据。——译注
② Merry White, *Coffee Life in Japan* (Berkeley, ca, 2012).
③ 郑永庆（Tei Ei-kei），郑成功胞弟田川七左卫门后裔郑永宁的养子。——译注

一切的权利并非切实可行的商业模式。郑永庆最终走向破产，在贫困中去世。

之后，日本制订了更多商业计划。老圣保罗咖啡馆是一家提供侍者服务的连锁咖啡馆，由水野龙在20世纪初开设。意大利不再向巴西输送劳工移民后，巴西制订了资助计划，引进了日本劳工，水野龙就曾受雇于巴西的咖啡种植园。这一时期，夏威夷的咖啡农场里也可见日本劳工移民的身影。与欧洲一样，日本在两次世界大战间隔年间也强调经济自给自足，大众咖啡文化发展因此受阻。

直到1960年，日本咖啡进口限令才被取消。次年，日本进口了25万袋咖啡豆。到了1990年，进口量增至533万袋。20世纪60年代中期，吃茶店①在日本社会开始风靡起来。相对较低的开设成本是吸引经营者的主要原因，而消费者数量则随日本经济繁荣而增加。到了1970年，日本已有5万家吃茶店，1982年增至16万家，达到了顶峰。此后，老式吃茶店与西式自助咖啡馆之间出现了分化。西式自助咖啡馆吸引着年轻消费者，比如罗多伦咖啡连锁店，该品牌在1980年开设了第一家分店，如今已有900多家分店。

许多吃茶店研发了自己独特的咖啡制作技术与程序，使用独特的网状过滤器或虹吸设备（最早由荷兰人引进）。这为哈里欧等玻璃器皿制造商创造了生产高规格、保温型

---

① 吃茶店（Kissaten），指日本出售咖啡、红茶等无酒精饮品以及和果子等简单食品的商店，而日本咖啡馆除了咖啡与轻食，也可提供酒类饮品和正餐。——译注

冲泡设备的机会。咖啡豆产地也逐渐开始受到重视，由于坦桑尼亚乞力马扎罗山看上去与日本神圣的富士山相似，因此产自该地的咖啡豆在日本颇受欢迎。

方便包装的新形式，特别是罐装的即饮饮品，推动了大众市场的发展。这些饮品最早由上岛咖啡公司生产，该公司于 1969 年推出了"上岛牛奶咖啡"。1973 年，上岛咖啡开始用自动售货机提供冷热咖啡，创造了一个巨大的、移动式"工业"咖啡市场。过了近 20 年，上岛咖啡才在自家产品系列中增加了无糖黑咖啡，这在某种程度上也说明了当时含糖饮品的流行。

日本销售的即饮罐装咖啡饮品

尽管如此，促使日本"国内"咖啡消费市场大幅增长的还是即溶咖啡；1983 年，每周消费的 8.5 杯咖啡中，有 5 杯属于即溶咖啡，这也令雀巢在日本咖啡市场遥遥领先。[1]

国际咖啡协议

全球咖啡贸易政治反映了消费者与生产者双方经济权力的转变。当美国食品连锁跨国公司为了进入海外市场而开始收购咖啡品牌时，一些国家内部出现了越来越多规模庞大、具有很强购买力的烘焙商。20世纪70年代，通用食品公司收购了瑞典烘焙商耶瓦利亚，莎莉公司收购了位于荷兰的杜威·埃格伯茨公司。1990年，菲利普·莫里斯公司旗下名单中新增了雅各布斯·祖哈德公司。而20世纪80年代，随着雀巢公司收购了希尔斯兄弟和蔡斯与桑伯恩等美国品牌，这一趋势又得以扭转。

罗布斯塔咖啡产量的增长从根本上改变了国际咖啡市场的供应方。由于世界咖啡供应量的增加，咖啡价格在1954年后下跌，导致拉丁美洲的生产商从1957年开始限制咖啡出口。不过，它们的干预措施未能成功，尽管巴西威胁要向市场倾销咖啡，但新的生产商并不在意，因为罗布斯塔的价格低于阿拉比卡。拒绝供应只会刺激买家转头购买罗布斯塔，因此到了1959年，巴西咖啡的库存约等于世界的全年咖啡消费量。

拉美国家开始游说美国达成一项全球协议，控制咖啡

---

[1] All Japan Coffee Association, 'Coffee Market in Japan', pdf document, www.coffee.ajca.or.jp/English, accessed 21 August 2017.

进出口。他们利用了 1959 年古巴革命引发的政治恐惧。一位哥伦比亚议员呼吁:"必须给我们的咖啡支付合理的价格,不然——上天保佑——群众会变成一支庞大的马克思主义革命军队,将我们统统扫入大海。"[1]在古巴导弹危机促使美国国会批准该协议之际,鉴于自身对巴西和哥伦比亚的依赖性,美国主要烘焙商认为支持这项协议是谨慎之举。欧洲的咖啡消费国也纷纷加入该协议,如此一来,仍归其统治的殖民地和新独立的生产国便有了经济保障。

1962 年,44 个出口国和 26 个进口国共同签署了《国际咖啡协定》(ICA)。其宗旨是:

保证以合理价格向消费者提供充足的咖啡,向生产者开放一定的咖啡市场,使生产与消费之间保持长期平衡,并在此基础上,实现供需的合理平衡。[2]

之后,在该协议的基础上成立了国际咖啡组织(ICO),总部设在伦敦。

国际咖啡组织委员会成为最高执行机构。提案通过需获得 70% 的生产者与消费者代表成员的投票。票数按成员出口量或进口量所占比例分配。巴西在 1000 张生产者票中拥有 346 张,而美国在 1000 张消费者票中占了 400 张。哥

---

[1] Gregory Dicum and Nina Luttinger, *The Coffee Book* (New York, 1999), p. 86.

[2] Richard Bilder, 'The International Coffee Agreement', *Law and Contemporary Problems*, XXVIII/2 (1963), p. 378.

伦比亚淡咖啡、其他淡咖啡、巴西自然咖啡和罗布斯塔咖啡，这四类咖啡的目标价格区间都已确定。成员对于每类咖啡都享有一定的出口配额。当实际价格超出规定的价格区间时，便放宽出口配额，从而降低市场价格，如 1975 年巴西遭遇毁灭性的霜冻后，便是以这种方式处理；如果实际价格低于规定的价格区间，那么为了重新抬高价格，则要收紧出口配额。从 1962 到 1989 年，运行的一直是这种配额制度。

全球咖啡产业链中的力量均势偏向了生产国，更确切地说，是在国际咖啡组织中代表这些国家的准政府机构。20 世纪 70 年代初，继布雷顿森林体系崩溃及石油价格冲击之后，消费者代表成员不愿意按配额执行，此时最大的咖啡生产者机构，如巴西咖啡管理局、哥伦比亚国家咖啡生产者协会和科特迪瓦的农产品价格稳定和支持银行，通过建立联合实体，在世界市场上买卖咖啡。它们利用自身预测库存量和采收量的专业"内部"知识，成功打击了金融投机者想方设法操纵期货价格的行为。

20 世纪 80 年代，配额制度依旧存在，主要出于政治的考量。1979 年尼加拉瓜发生桑地诺革命后，美国里根政府希望在萨尔瓦多、危地马拉、尼加拉瓜和哥伦比亚的内战中防止左翼进一步取得胜利。这些国家爆发内战，某种程度上是因为自身咖啡种植产业内部财富分配不均。在危地马拉,1%的咖啡农场产出了全国 56% 的咖啡产量。[1]而在萨尔瓦多,在咖啡农场劳作的土著农民沦为军事统治者用暴力进行种族压迫的目标。另一方面,游击队要求中产阶级农场主缴纳"战争税",否则将烧毁他们的建筑,抢占其土地。在桑地诺的统治下,尼加拉瓜咖啡机构 ENCAFE 给生产商的收购价只有咖啡出口价格的 10%。[2]

巴西受霜冻所害的咖啡植物。1975 年,"黑霜"摧毁了巴西 50 多万棵树。次年,咖啡产量下降了 60%,而咖啡生豆的价格在 1975 至 1977 年之间涨了两倍。

相较于其他商品,配额制度可以带来相对安全的收益,鼓励了菲律宾和印度尼西亚等生产国提高咖啡产量。《国际

① Steven Topik, John M. Talbot and Mario Samper, 'Globalization, Neoliberalism, and the Latin American Coffee Societies', *Latin American Perspectives*, XXXVII/2 (2010), p. 12.
② Mark Prendergast, *Uncommon Grounds* (New York, 2010), p. 317.

咖啡协定》规定的配额也起到额外的激励作用，使"实际生产力"与"以往"世界市场份额达到平衡。

新生产的咖啡大多属于罗布斯塔咖啡，切合了速溶咖啡产品的需求。进口商十分希望利用这些价格低廉的新品种咖啡供应，于是研发出了"清洗咖啡豆"的方法，用蒸汽处理豆子以减轻其苦味。到1976年，雀巢公司已在21个生产国成立咖啡处理的分支机构。[1]有些国家建立起了自己的咖啡产业。为了出口咖啡粉，厄瓜多尔选择种植及处理罗布斯塔咖啡豆。巴西在圣埃斯皮里图州引进了罗布斯塔的变种——柯林隆咖啡豆，建设了处理豆子的一系列基础设施。速溶咖啡在巴西人中流行起来，如今，罗布斯塔咖啡大约占巴西咖啡总产量的20%。

一些中美洲政府机构利用稳定局势，大力投资农业研究以提高产量。所谓的"技术化"，包括引进能在全日照条件下生长的矮小品种，以及采用化学肥料，使生产者的产量大幅提高：在20世纪70年代中期和90年代初，哥伦比亚的采收量增加了54%，哥斯达黎加增加了89%，洪都拉斯增加了140%。[2]不过，对于产出的过剩的咖啡，出口商并没有选择销毁，而是将之廉价投放到《国际咖啡协定》未覆盖到的市场，如"苏联集团"[3]。1989年，哥斯达黎加40%的咖啡都是以半价或更低的价格出售，其中一些还

---

① John M. Talbot, *Grounds for Agreement* (Lanham, MD, 2004), p. 61.

② 同上，pp. 77–81。

③ 指"二战"后，西方国家对以苏联为首的社会主义阵营的称呼。——译注

被用于与捷克斯洛伐克公共汽车等商品进行交换。

烘焙商渴求有新的咖啡供应，于是往往采用通过非配额国家进入配额国家的"游客"咖啡。若再在接收港动些手脚，可能将高级咖啡与低级咖啡互调，前者原本指定给非成员国，而后者则打算出售给成员国，如此一来，各方都以低于配额的价格获得了咖啡。这种现象一直存在，因为成员国都不愿为了反映需求变化而自掏腰包来调整配额。

1989 年 9 月，随着苏联和桑地诺政权的衰落，美国中断了对配额制度的支持，最终于 1993 年退出国际咖啡组织。虽然美国是唯一一个选择退出的消费国，但缺少它，整个监管体系都无法实施。与此同时，生产国之间的利益冲突导致几乎没有国家想要继续维持下去。一些政府机构被解散，特别是巴西咖啡管理局。今天，国际咖啡组织作为各国的信息交流机构仍在运作，不过不再发挥对全球咖啡供应链的管理作用。

尽管功能失调，配额制度却使咖啡市场保持在相对稳定的状态。在实行配额制度的最后八年里，咖啡每月的指标价格变化率为 14.8%；而在那之后的八年，变化率达到37%。1984 至 1988 年，咖啡平均指标价格为 1.34 美元 / 磅；1989 至 1993 年，由于咖啡市场上供应过多，指标价格下降到 0.77 美元 / 磅。[1]不过，巴西霜冻导致了供应量下降，阻止了价格进一步下跌，但全球咖啡贸易体系显然又变得

---

① International Coffee Organization, *World Coffee Trade (1963–2013): A Review of the Markets, Challenges and Opportunities Facing the Sector* (London, 2014).

不稳定。

越　南

　　鉴于《国际咖啡协定》的终止，越南趁此在 20 世纪的最后十年彻底改变了世界咖啡贸易。1999 年，它超过了哥伦比亚，成为世界第二大咖啡生产国，而在 1988 年，它还仅排在第 22 位。越南的成功离不开罗布斯塔咖啡树的种植，越南因此成为世界最大的罗布斯塔咖啡出口国。

　　19 世纪 50 年代，传教士便在越南种植少量的阿拉比卡咖啡树，但直到法国殖民统治时期，咖啡仍只是一种小规模种植的作物。到了 1975 年，共产党领导下的北方和美国支持的南方之间的长久战结束时，越南只余下 60 公顷（约

越南，1994 年。越南中部高地邦美蜀附近的农民正在称晒干的罗布斯塔咖啡樱桃。

合 148 英亩）的咖啡种植区。越共政权取得胜利后，试图稳定从前受南方控制的地区。越南民主共和国的忠心耿耿的农民被政府鼓励迁移到中部高地地区，通过结合现有土地的国有化和积极的森林砍伐计划，在那里建立国有农场和农业合作社。政府鼓励他们在此种植咖啡树，将咖啡出口到越南的盟友"苏联集团"。

直到 20 世纪 80 年代，越南政府开始经济改革后，咖啡产量才开始飙升。20 世纪 90 年代，政府开始将土地向私人转移，这意味着到 2000 年，越南 90% 的咖啡生产都来自小农户，而他们耕种的土地不到 1 公顷（约合 2.5 英亩）。国家依然大力支持他们，继续通过土地补贴、金融信贷和技术援助（如利用化肥）来激励生产。因此，在 1988 至 1999年间，越南咖啡产量显著增长，年均增长率为 24%，平均产量远高于竞争国。

1995 年，威拿咖啡——负责越南咖啡产业，包括开发、营销和出口活动的国家机构——摇身一变，成为越南国家咖啡总公司。它管理越南余下所有国有农场以及大大小小的加工、贸易和服务商，控制该国大约 40% 的咖啡出口量，并经营两大速溶咖啡集团之一（另一个属于雀巢公司）。

越南咖啡生产的迅速扩张似乎是为了在"苏联集团"解体后，巩固自身政权的政治地位。越南在提高出口收入的同时，还允许农民直接接触市场，实现"发家致富"。不过，这也具有一定的危险性：供应过剩，最终会导致咖啡价格下跌——正如 1998 年后所发生的那样。

咖啡危机

1998 年，国际咖啡组织的咖啡综合指标价格为 109 美分 / 磅；到了 2002 年，价格已跌至不到 48 美分。之后虽然价格也有上涨，但直到 2007 年才恢复到 100 至 107 美分的价格区间。这种长期性的价格暴跌影响极大：生产者陷入贫困，整个咖啡行业在消费国的公共形象受损。

这个问题的关键在于持续性的供大于求。2001 至 2002 年，世界咖啡产量为 1.13 亿袋，另外还有 4000 万袋库存，但消费量为 1.06 亿袋。国际咖啡组织执行董事称："咖啡过剩现象的根源在于越南咖啡生产和巴西新种植园的迅速扩张。"①

价格下跌对不同地方的咖啡行业造成的影响有所不同。对于巴西等生产成本低、技术发达、汇率波动良好的地方，仍有机会获利。反之，在大多数非洲国家、一些中美洲国家以及许多亚洲国家，自给自足的农民将咖啡作为经济作物种植，导致用于医药、教育、食品或偿债的资金减少。

危地马拉从事咖啡业的劳动力减半。哥伦比亚农民砍掉自家咖啡树，转而种植用于毒品贸易的古柯树。墨西哥种植者中有许多人放弃种植咖啡树，还试图非法前往美国，但往往在这个过程中丧生。随着政治冲突愈演愈烈，墨西

---

① Néstor Osorio, 'The Global Coffee Crisis: A Threat to Sustainable Development', Submission to World Summit on Sustainable Development (Johannesburg, 2002).

哥咖啡生产中心恰帕斯的农民支持萨帕塔游击队反抗政府的运动。

甚至在越南，有些农民为了还债，被迫变卖家产。中部高地的贫困水平达到50%，其中30%的人口正遭受饥饿与营养不良。罗布斯塔咖啡的价格从1998年的83美分开始不断下跌，到了2001年，仅为28美分。这给乌干达等严重依赖咖啡出口的国家带来了沉重打击。

然而，越南的咖啡产量仍在继续增长，因为农民试图通过生产来摆脱危机。1990至1991年，该国的咖啡产量为130万袋；2000至2001年，为1480万袋；到了2015至2016年，则惊人地增至2870万袋，超过了整个非洲大陆的产量。其他亚洲国家也纷纷效仿越南，缅甸、老挝和泰国三个东南亚国家都将咖啡发展成了重要产业，印度尼西亚和印度也跻身咖啡生产国前六名的行列。而无一例外，90%以上的产量都来自罗布斯塔咖啡。

2010年，破坏性极强的咖啡叶锈病暴发，开始在整个拉丁美洲蔓延，之后咖啡价格加速上涨。这场冲击迫使咖啡的供求关系重新达到平衡，特别是对于高品质的阿拉比卡咖啡而言，这也导致咖啡综合价格整整十年都保持在每磅120美分以上。但要牢记，这种新获得的稳定性在某种程度上牺牲了那些因价格下跌、气候干旱或植物病害而被迫离开农业的人。

1989年配额制度走向崩溃后，咖啡价格的波动证实了解除管制所带来的诸多危险。甚至叶锈病的暴发也被归咎于准政府机构的瓦解，它们本是国家研究和应对作物病害的协调者。[1]不过，尽管配额制度调控了咖啡流入市场，但它偏向既定的生产商，政府机构通常无法让咖啡的利润回到农民手中。

矛盾的是，伴随着咖啡危机，同时爆发了所谓的"拿铁革命"，其主要特征是收费高昂的咖啡店迅速扩张，这招来了有些人对于"在咖啡杯中消费贫穷"的批评。[2]然而，其他人将这种新现象视作商机，认为可以将咖啡重新打造为"精品饮品"，使之去商品化，从而在整个价值链中获取更大的收益。

---

[1] Stuart McCook and John Vandermeer, 'The Big Rust and the Red Queen', *Phytopathology Review*, CV (2015), pp. 1164–73.

[2] Oxfam, *Mugged: Poverty in Your Coffee Cup* (Oxford, 2002).

精品饮品

20世纪末,咖啡被重新定位为精品饮品,对全球咖啡业产生了深远影响。起初是美国独立烘焙商抗议咖啡商品化,后来咖啡产业的集中化又催生了国际咖啡连锁店的扩张、"第三波"咖啡浪潮的兴起、咖啡胶囊的发展,以及一连串关于合理消费咖啡的热议。可以说,精品咖啡在刺激非传统市场消费方面的影响,为开辟咖啡历史的新纪元奠定了基础。

精品咖啡的诞生

1958到1978年,美国四大烘焙商所占市场份额从46%增至69%。到2000年,"三大巨头"——宝洁、卡夫和莎莉——控制了美国超过80%的咖啡零售市场。它们纷纷展开价格竞争:降低拼配对象成本,有些品牌打出的广告都是自家产品可以用更少的量冲泡出同等风味的咖啡。

然而,这些策略终究未能将美国人均咖啡消费量稳步下降的状况扭转过来,美国人均咖啡消费量从1960年的7.25千克(约合16磅)降至1995年的2.7千克(约合6磅),尽管自动咖啡机在20世纪70年代便让美国人(尤其男性)开始改用滴滤式冲泡方式。相比之下,人们对于含咖啡因的软饮消费高涨,原因五花八门,其中包括中央供暖系统的普及、快餐店的兴起,以及朝气蓬勃的广告的吸引。

独立烘焙商的数量同样在下降,1945年还有大约1500家,到了1972年便只剩162家。为了生存下去,他们发展出了另一种商业策略。烘焙商们决定，与其进行价格竞争，不如展开品质较量，这样能提高咖啡豆的利润率。该策略适用于消费经济，不同社会群体早已开始利用购买力来彰显自己的生活方式、价值观及品位，还可能展现自己见多识广或财力雄厚；坚持"另类"，反企业价值观；或对"真正的"手工物的偏爱。

1977年，西雅图一家早期的精品咖啡馆，出售供家庭消费的咖啡豆，此处未见浓缩咖啡机的身影。这家店的名字是……星巴克。

咖啡是20世纪60年代美国反主流文化的重要组成部分，而旧金山正是该文化的精神家园所在。嬉皮士们常出没于北滩的浓缩咖啡吧，这里的经营者都是意大利移民，他们会从伯克利的阿尔弗雷德·皮特商店购买咖啡豆。皮特乃荷兰人，相较于普通的"一杯咖啡"，他家的咖啡烘焙程度较深，冲泡风味也更香。尽管皮特几乎毫不掩饰对许多顾客的轻蔑，但皮爷咖啡馆仍成为人们一心想要体验"欧洲"咖啡的圣地。

埃尔娜·克努森是首位使用"精品咖啡"一词的人。刚开始，她担任与旧金山咖啡进口商对接的秘书，20世纪70年代中期，她说服这些人让她尝试销售小批量的优质咖啡。她为新一代独立烘焙商找到了合适的定位，当时许多人早已"脱离了"传统的职业道路。

1982年，一群烘焙商成立了美国精品咖啡协会（SCAA），将"精品"的概念定义为咖啡杯里的独特口感。其产品范围包括高级出口咖啡，如肯尼亚AA级咖啡，以及类似于瑞士杏仁摩卡等拼配咖啡和调味咖啡——在今天，这些都不太可能被归到精品咖啡的行列。它们当时的销售场所都是雅皮士喜欢光顾的美食熟食店，这群人是年轻的城市职业者，其购买力不断增长，而这构成了20世纪80年代一系列美食家革命的基础。

当咖啡行业的重心从销售咖啡豆转向提供咖啡饮品时，精品咖啡立即蓬勃发展起来。而西雅图正处于发展的中心位置：1980年，该市出现了首批装有浓缩咖啡机的咖啡车；到了1990年，单轨列车站、渡轮码头和主要商店附近有200多辆咖啡车。显然，比起办公室提供的免费咖啡，上班族们更愿意花钱消费外带的精品饮品。如今，西雅图只剩下一两辆咖啡车，其他都被由这座城市扩及全世界的咖啡店革命扫除了。

星巴克的起源

1971 年，三位大学同窗好友共同创立了星巴克。它在当时主要销售来自阿尔弗雷德·皮特商店供应的咖啡豆，三位创始人之后也延续了皮特咖啡豆深度烘焙的风味。霍华德·舒尔茨原是布鲁克林一家公司的推销人员，该公司是星巴克的设备供应商之一。1982 年，他拜访并说服几位创始人聘请他担任销售与营销总监。1983 年，舒尔茨访问米兰，他在此

> 找到了驱动我自己人生与星巴克发展的灵感与愿景……若我们能在美国重建真正的意大利咖啡吧文化……星巴克可以成为一种美好的体验，而不仅仅是一家出色的零售店。[①]

不过他未能成功说服星巴克的几位老板，于是离开了这家公司，在 1986 年开了家名为 Il Giornale 的咖啡馆。他选择这个名字，是以为它指的是"每日"，即意大利人去当地咖啡吧的频率。但事实上，它的意思是"报纸"。

美国还有其他很多与舒尔茨的愿景相悖的地方。这里的顾客不愿站在吧台前喝咖啡，更倾向于坐在桌前聊天。相较于瓷杯，他们更喜欢纸杯，因为这样就可以把饮料带

---

① Howard Schultz and Dori Jones Lang, *Pour Your Heart Into It* (New York, 1997), p. 52.

回工作的地方喝。而作为背景音乐的歌剧以及戴着蝴蝶领结的咖啡师，则与太平洋西北部随意的氛围格格不入。

但舒尔茨的一番改进让美国顾客体验到了他们想要享受的"意式"咖啡服务，他一下子便获得了成功。1987年，他将这种经营模式带去了星巴克，当这家公司最后一位最初的创办人离开，去旧金山接手皮爷咖啡时，他将之一举收购。

咖啡馆的经营模式

咖啡馆的经营模式融合了两大元素：咖啡与环境。前者要为后者买单。

事实证明，意式咖啡是向美国消费者推出精品咖啡的最佳选择，通过牛奶的甜味，仍可以分辨出浓缩咖啡在口中的独特风味。拿铁咖啡是最受欢迎的，因为与卡布奇诺相比，拿铁用的是蒸牛奶而非起泡牛奶，喝起来的口感更加绵密甜醇。调味糖浆的加入使咖啡馆能够推出定制系列，提供季节性饮品，如蛋奶酒拿铁。事实证明，平价确实比正宗更为重要：星巴克标准中杯卡布奇诺的量是意大利卡布奇诺的两倍，甜度也更高。

到了1994年，美国精品咖啡馆以浓缩咖啡为主的饮品在销量上超过了冲泡咖啡。咖啡师"手工制作"饮品的情景——研磨新鲜咖啡豆，从咖啡机里取出咖啡，打出奶泡并倒入牛奶，在上面撒上肉桂、巧克力和（或）巧克力粉——都让咖啡制作过程中的附加值变得显而易见。因此，消费者愿意花高价购买他们在家里享受不到的优质产品。

这其中的高利润包含了消费者为享受咖啡时所置身的舒适环境而支付的费用。沙发、音乐、报纸以及带有给婴儿换尿布设施的干净厕所都能促成一笔"20分钟的交易"。消费者享用咖啡馆提供的这些服务设施需要支付所谓的租借费，这就包含在咖啡的价格里。咖啡馆让人感到平等，因为会按照顾客到来的先后顺序提供服务；同时具有包容性，因为主要经营的是咖啡而非酒，为女性、儿童和不饮酒者提供了"安全"空间。

舒尔茨鼓吹星巴克是工作与家庭之间"第三场所"的典范，据社会学家雷·奥尔登堡的描述，在星巴克，不相关的人之间轻松随意的接触创造了一种群体感。[①]而行为学研究发现，所谓陌生人之间会彼此聊天的说法几乎毫无根据：咖啡馆的吸引力恰恰在于，虽然置身人群，但不一定要融入。数字技术的不断进步——笔记本电脑、手机、无线网——让个人在"消费"咖啡馆氛围的同时，可以继续工作或在社交媒体上聊天。

星巴克连锁店的霸主形象

事实证明，舒尔茨擅长为咖啡馆扩张融资。1992年，星巴克首次公开上市后，集中收购了一些人流量大的店铺，

---

① Ray Oldenburg, *The Great Good Place* (New York, 1989).

选址通常在同一条街道彼此相邻的位置。这带动了星巴克整体贸易的增长，因为人们不会去太偏离自己日常活动范围的地方喝咖啡。不过，一旦改喝精品咖啡，他们无论在哪儿都倾向于喝这种咖啡，于是不仅促进了星巴克贸易的增长，也推动了整个行业的发展。

品牌建设对于星巴克维持自身高品质的形象至关重要。为了保证消费者在任何一家分店的服务体验相同，星巴克员工必须遵循消费者服务手册，不断冲泡一模一样的饮品，因此瑞士超级自动按钮式浓缩咖啡机在 1999 年取代了传统的意大利设备。它还给名人付费，"发现"他们正在喝星巴克外带咖啡，然后拍下照片。星巴克维持了自身在咖啡馆领域的霸主形象，它稳居主导地位，成功定义了消费者所理解的咖啡馆这个概念。

美国精品咖啡馆与星巴克分店[1]

年份	精品咖啡馆	星巴克分店	百分比
1989	585	46	7.9
1994	3,600	425	11.8
2000	12,600	2,776	22.0
2013	29,308	11,962	40.8

根据 2016 年的一项报道，当美国人被问及前一天喝咖啡的情况时，喝"极品咖啡"或精品咖啡的人与喝普通咖啡的人一样多。[2]这是由于自 2008 年以来，以浓缩咖啡为

---

[1] Data available at www.sca.org and www.statista.com.

[2] 'What Are We Drinking? Understanding Coffee Consumption Trends', www.nationalcoffeeblog.org, 2016.

主的饮品消费增长了三倍，也同时反映了日常快餐连锁店和街角小店（如唐恩都乐和麦当劳）都在销售这类咖啡。如今和"一杯咖啡"一样，拿铁咖啡已成为美国咖啡的特色。

2017年，西雅图的星巴克咖啡烘焙工坊。以现场烘焙、放大的现实体验感、大量咖啡豆以及各种各样的冲泡方式为特色，咖啡烘焙工坊旨在凸显星巴克这一品牌在"第三波"咖啡文化中的身份。

国际化

国际化是星巴克战略的另一重要组成部分——2017年1月1日，星巴克的运营门店多达25,734家，遍及75个国家。1996年，在日本和新加坡，第一家星巴克连锁店开业后，其国际化战略迅速扩张到东南亚"四小虎"[①]中。咖啡馆文化受到这些国家中产阶级年轻群体的热烈欢迎，他们热衷于追赶美国潮流，将咖啡馆作为社交与学习的场所。

咖啡馆的概念在欧洲的普及往往早于星巴克，追随者和移民为满足当地消费者的口味，对其进行了一番调整。20世纪80年代末，在英国经营意式咖啡馆的伦敦咖啡烘焙商咖世家开始经营浓缩咖啡吧。它于1995年被啤酒与休闲集团惠特布雷德收购，该公司很有先见之明，预料到咖啡馆将取代酒吧成为英国社交中心。到2017年，咖世家门店已从1995年的41家扩张至2100家，成为最大的咖啡馆经营商。与英国其他所有连锁店一样，咖世家聘请的咖啡师大都来自国外：年轻人颇会利用当时欧盟内部自由往来的法律。[②]

品牌连锁店促进了英国咖啡馆的发展，而自21世纪10年代以来，百货公司与花园中心连锁店成为发展势头最为迅猛的领域，这在一定程度上也反映了意式咖啡已成为主流的英国饮品。酒吧数量不断下降，如今开始在白天供应咖啡，以求生存。

连锁咖啡馆已经遍布整个欧洲大陆，但具体的发展情况还是取决于当地咖啡文化的特征。作为时尚买手的瓦妮莎·库尔曼在纽约体验过咖啡馆的服务后，于1998年创立

[①] 东南亚"四小虎"，指印度尼西亚、泰国、马来西亚和菲律宾四国。——译注

[②] Jonathan Morris, 'Why Espresso? Explaining Changes in European Coffee Preferences', *European Review of History*, XX/5 (2013), pp. 881–901.

了德国第一家咖啡连锁店巴尔扎克咖啡。高级浓缩咖啡饮品如今大约占德国除家庭外市场一半的份额，不过大多由连锁面包店供应。在法国，浓缩咖啡饮品的发展颇为成熟，连锁咖啡馆最早于20世纪90年代出现，但直到2008年经济衰退后才真正开始快速发展，代替小酒馆为消费者提供快速服务。希腊的发展情况也同样如此。

作为浓缩咖啡的发源地，意大利始终是这场精品咖啡变革的主要受益者。烘焙咖啡出口量从1988年的0.12亿公斤增至2015年的1.71亿多公斤。意大利公司控制了全球商业浓缩咖啡机市场份额的70%，其产量的常规出口量超过90%。意利和塞加弗雷多等咖啡公司利用商标注册与特许经营，在世界各地建立自己的品牌连锁店。但意大利当地并

浓缩咖啡马提尼：20世纪80年代，伦敦酒保迪克·布拉德肖在一位女顾客提出想要喝到"能唤醒我，让我兴奋起来"的饮品时，调制出了这款鸡尾酒。

一台意大利制造的浓缩咖啡机正在制作咖啡。意大利国家咖啡协会（INEI）会对咖啡师、拼配咖啡以及咖啡设备进行认证，确保其能够制作出符合合法注册标准的咖啡，见图中机器上的标志。

未出现任何咖啡连锁店，因为这里的浓缩咖啡价格基本不可能溢价。星巴克直到 2018 年才在米兰开业，此时距离舒尔茨那次受到启发的访问已大约过去了 35 年，而它所采用的形式主要是为了凸显自身在"第三波"咖啡文化中的身份。

"第三波"咖啡浪潮

2000 年，蒂莫西·卡斯尔首度使用"'第三波'咖啡浪潮"这一术语；2003 年，美国烘焙商特里西·罗思格柏在一篇颇有影响力的文章中提到该词，使其普及。[1]她写道，"第一波"面向大众市场的烘焙商"让劣质咖啡变得普遍"。精品咖啡经营者最初"通过小型烘焙业务开始面向指定商店……提供浓缩咖啡"，但这种形式被星巴克等"第二波"咖啡巨头压住了风头，后者"想让精品咖啡自动化或同质化"。"第三波"咖啡浪潮将追求以一种"无规则"的方式来制作出色的咖啡。

咖啡师大赛是"第三波"咖啡文化的核心。2000年，第一届世界咖啡师大赛在摩纳哥举行。参赛者需在15分钟内准备一组四杯的浓缩咖啡、卡布奇诺和"创意饮品"，由评委根据其冲泡技术、展示技巧以及饮品的感官品质进行评判。设备制造商争相让自家机器可以被划入符合比赛标准的行列。烘焙商纷纷全职培训咖啡师，让他们使用特别采购的拼配咖啡参加比赛。获胜者将一举成名，获得高薪的咨询与代言合同。

在"第三波"咖啡浪潮中，咖啡师们将影响浓缩咖啡制作与风味的既有参数试验了个遍，由此摆脱了意式咖啡传统。这一系列试验也带来了一些新研发的饮品，比如馥芮白，即在数份浓缩咖啡中加入丝绒般微微发泡的牛奶，最后再制作拿铁拉花——这一切都要求咖啡师具有精湛的制作技术。2007年，馥芮白被澳大利亚咖啡师带去了伦敦，

① Timothy J. Castle and Christopher M. Lee, 'The Coming Third Wave of Coffee Shops', *Tea and Coffee Asia* (December 1999–February 2000), p. 14; Trish Rothgeb Skeie, 'Norway and Coffee', *The Flamekeeper: Newsletter of the Roasters Guild, SCAA* (Spring 2003).

而到 2010 年，它已经进入伦敦主流咖啡连锁店的菜单，后来又跨越了大西洋。

韩国首尔。来自英国的戴尔·哈里斯获得了 2017 年世界咖啡师大赛冠军，图上是他正在参加这场即将夺魁的比赛。

"第三波"咖啡馆往往都是小本经营，鼓励经营者的更多是对咖啡馆的热情，而非其盈利能力。咖啡馆内饰简洁，只有基本的桌椅，反衬出柜台上那些几乎占据了所有投资的高科技机器。这种互不相容的氛围格外明显，也十分常见，已成为"第三波"咖啡自身的品牌形象。

"第三波"咖啡烘焙商会从同一地区采购单品咖啡，最好能溯源到同一农场或同一生产者合作商。他们采取了工匠式运作——小批量烘焙，随时调整风味，力求每个批次的产品都能达到最佳水准。通常采用浅烘焙——主要是为了带出咖啡豆自身的风味特点，而不是让人仅仅关注烘焙

20世纪80年代，馥芮白在澳大利亚逐渐演变，之所以取这个名字，很可能是为了将它表面丝绒般的牛奶与卡布奇诺覆盖的那层奶泡区分开来。当咖啡馆难以打出鲜奶泡时，就会贴出"仅供馥芮白"的告示。

特色本身。

1999年，美国主要的精品咖啡购买商开始在咖啡生产国组织"卓越杯"咖啡豆竞赛。农民提交自己生产的咖啡豆，由杯测师组成的国际评审团进行评估，获奖的咖啡豆将在网上以天价拍卖。在2016年的拍卖会上，获奖的咖啡豆分别来自日本、韩国、中国台湾、保加利亚、澳大利亚、荷兰和美国，这也反映了精品咖啡的全球传播。

"第三波"咖啡浪潮逐渐将目光从浓缩咖啡移至用其他方式冲泡的咖啡上，后者最能体现单品精品咖啡等级的微妙性。日本研发的咖啡设备——如哈里欧V60过滤器和虹吸壶——在"第三波"浪潮咖啡馆中已变得十分常见。20世纪40年代首次生产的凯梅克斯咖啡设备逐渐受到欢迎，因为它改进后的纸质过滤器过滤出的咖啡格外澄净，品相看上去更接近于茶。

这波咖啡浪潮被贴切地形容为一种跨国"亚文化"形式，将自身理念、标志性品牌、咖啡爱好者的杂志式出版物以及有重要影响力的人杂糅在一起。互联网让一切成为可能，小烘焙商可以在全国各地找到客户——所谓的"产消者"——可以讨论定制咖啡机器的最佳方案，而鉴赏家们则可以在网上阅读到最新的咖啡评论。这些不同的群体又可以在咖啡节上聚在一起，比如自 2011 年以来在伦敦举办的咖啡节。

单杯咖啡

精品咖啡革命令人产生了在家制作类似饮品的冲动。而采用"单杯"咖啡胶囊的机器使之成为现实。研磨好的部分咖啡被密封在铝制胶囊中以保持新鲜度。将胶囊置于制作咖啡的机器里，用多针刺穿胶囊顶部，然后注入热水，使胶囊在压力下破裂，输出咖啡。尽管这套制作流程兼具便利性与清洁性，但这些胶囊只能由专门的设备回收，需要消费者承诺收集并返还用过的胶囊，而不是将之归为家庭垃圾。

雀巢公司 1986 年成立的品牌奈斯派索率先采用了这套制作浓缩咖啡式饮品的技术，并一直保持着该领域全球领导者的地位。在美国市场上，绿山咖啡烘焙公司 1998 年率先推出了克里格 K 杯，复刻美式滴滤咖啡，进而主导着该市场。

奈斯派索专为小餐馆、航空公司和铁路公司等提供服务的经营商开发产品，满足它们对咖啡机的需求，这样既不用招聘经过培训的咖啡师，又能节省大量空间。克里格则面向写字楼及酒店客房市场，因为胶囊可避免随意丢弃和浪费。而很快便能看出，这些优点让这种咖啡制作方式格外吸引家庭消费者。

奈斯派索对自身产品的定位是：提供进入美妙咖啡世界的入场券。除了适合各种口味及偏好的浓缩咖啡拼配，它还向自家会员提供所谓的特级和限量款咖啡。2000 年，奈斯派索在巴黎开设了首家零售精品实体店；到 2015 年年底，

俄罗斯圣彼得堡奈斯派索精品实体店的某个区域，展示了店里的咖啡机与胶囊，2017 年。

它已在60个国家有467家店面，占据了各大城市的黄金地段——靠近奢侈品牌店的地方。奈斯派索的各种品牌联合战略，如推出保时捷设计的咖啡机，巩固了它作为高级生活方式产品的地位，自2005年以来，还一直邀请乔治·克鲁尼担任其主要品牌大使。

2000至2010年，奈斯派索的年增长率超过30%。2010年，由于其产品溢价，瑞士在出口额上成为全球最大的烘焙咖啡出口国，不过其出口量仅排在第五位。

2012年，保护奈斯派索与克里格专有技术的专利到期。之后五年间，单杯咖啡的全球市场至少增长了50%。连锁咖啡馆针对家庭推出了一系列产品，充分利用自身品牌，其他制造商试图生产兼容款胶囊，以比奈斯派索和克里格更低的价格出售产品，与之展开竞争，而专业经营者则进一步发掘胶囊在"第三波"咖啡浪潮中的潜力。到2017年，美国和英国至少三分之一的咖啡饮用者使用胶囊机制作咖啡。

彰显道德的咖啡

在咖啡行业采取的认证体系中，精品咖啡起着主导作用，经过认证的标签意味着特定咖啡的供应链在环境或社会经济方面具有可持续性。环境认证包括"有机""鸟类友善"和"雨林联盟"，这些标签有利于改进可持续性的耕作技术，促进生物多样性。

首个社会认证计划由"公平贸易"运动发展而来。1988年，荷兰宗教组织禾众基金会创立了名为马克斯·哈韦拉尔的标签——以那本谴责爪哇咖啡殖民贸易的同名小说命名。该基金会开始从生产商合作社采购咖啡，最初是在墨西哥，然后在德国和荷兰进行销售。1989年，乐施会等英国慈善机构有样学样，创立了"咖啡直达"这个品牌，通过教堂大厅和慈善商店销售。1997年，国际公平贸易协会成立，将各国认证体系结合起来。

如今，"公平贸易"协会仍是保证生产商咖啡最低价格的唯一认证体系。从定价体系便能看出该咖啡属于阿拉比卡还是罗布斯塔，日晒还是水洗，有机还是无机。此外，出口合作社的产品还会获得一定程度的社会溢价，用于改善咖啡种植区的生活环境。自2011年以来，这部分溢价的四分之一都须用于提高咖啡品质。若世界市场价格超过交货时的"公平贸易"协会定价，则说明它可以将价格定得更高。

"公平贸易"各个协会自身并不购买或出售咖啡，而是准许产品被贴上"公平贸易"协会的标签，并收取相应费用。为此，商品链中的所有参与者都必须经过认证，以确保他们遵守"公平贸易"协会的标准。有人批评，"公平贸易"协会的价格保证虽然保护了生产商，使其不受市场影响，且无需对市场作出反应，但同时也导致惰性的产生。此外，大部分差价仍由发达国家的消费者支付，要么是付给烘焙商，要么是支持认证体系的运营成本。

2013 年，卢布尔雅那开设了斯洛文尼亚首家"公平贸易"协会商店，出售带有"公平贸易"协会认证标签的有机咖啡。这款咖啡由 EZA 提供，EZA 属于奥地利社会企业，由天主教社会机构成立。

　　若分析"公平贸易"协会认证体系对生产商的影响，结果则喜忧参半。在 21 世纪头一个十年中期爆发的那场咖啡危机中，底价是重要的安全保障，许多群体纷纷受益于社会溢价的再投资。自此，咖啡的"公平贸易"协会价格与市场价格之间保持相对较小的差距。拉丁美洲研究表明，这之间的差距并不总是足以抵消农民由于需要向采摘者支付更高的工资而造成的潜在收入损失，而有机咖啡所带来的溢价并不能弥补因转而使用该体系而导致的产量减少。[1]2017 年的一项研究表明，尽管亚洲"公平贸易"协

---

① Daniel Jaffee, *Brewing Justice* (Berkeley, ca, 2007); Tina Beuchelt and Manfred Zeller, 'Profits and Poverty', *Ecological Economics*, LXX (2011), pp. 1316–24.

会生产商的收入足以用来养家糊口，但非洲的生产商却做不到，因为他们的资产太少，所以无法获得咖啡溢价所带来的收益。①

对于烘焙商和经营者而言，"公平贸易"协会的标签价值在于彰显自身的道德信念，让自己能够收取足以支付额外采购成本的溢价。在咖啡危机时期，由于媒体经常强调高价拿铁与饥饿咖啡农之间的矛盾，烘焙商和经营者对此格外敏感。这就导致星巴克在1999年西雅图的反全球化骚乱中成为众矢之的，因为尽管它拥有大量的商品烘焙商，但它向供应商支付的费用却是最少的。

然而，市面上能够买到的"公平贸易"协会认证的咖啡相对较少，因为该协会坚持要通过生产商合作社采购。于是，这首先排除了不属于合作社的大型种植园、独立农户以及小农户的咖啡，还有大量被最初的认证费用挡在门外的合作社。

尽管这些生产商也可以采取其他认证体系，但终究是由贸易商和生产商决定标签的溢价。由大型烘焙商和生产国共同制定的"咖啡社区共同准则"（Common Code for the Coffee Community，简称4C），在2007年推出了一组关于社会、环境以及经济标准的基线。4C生产商认证的价格按产量分级，同样在许多小农户的承受范围内，而这些容易达到的标准，也让它能吸引到从多个供应商那里采购的跨

① True Price, *Assessing Coffee Farmer Income* (Amsterdam, 2017).

国公司。

2012 年，美国"公平贸易"协会脱离了国际"公平贸易"协会，从而能够开始对非合作组织的生产商进行认证。它认为，这更有利于保护劳动者与独立小农户的利益，同时让更多的消费者有机会选择购买"公平贸易"协会认证的产品。

到 2013 年，全球生产的大约 40% 的咖啡都符合某类认证标准。[①]有咖啡爱好者认为，这代表了让全球资本主义承担社会责任所取得的最伟大的胜利之一。也有批评者表示，这是公共关系的胜利，使咖啡业在参与"释放美德信号"的同时利用消费者的道德关切赚钱。

"第三波"咖啡浪潮的烘焙商们反对"公平贸易"协会对咖啡品质以及可追溯性的不重视。对于美国"第三波"咖啡浪潮的主要烘焙商树墩城、知识分子和反文化，其买家发展出了另一种"直接贸易"的模式：识别出潜在的优秀咖啡种植者，与之合作以保证咖啡品质，然后按反映品质的价格直接从他们那里购买，这样的价格远高于"公平贸易"协会所能达成的价格。这种伙伴关系会对种植者产生变革般的影响，不过仅限于当地有条件种植精品咖啡树的农民。

将"公平贸易"协会、直接贸易及其他发展计划联系起来的干预措施，为改善某些国家的咖啡生产条件做出了巨大贡献。卢旺达以商品咖啡生产为中心的咖啡基础设施，

---

① David Levy et al., 'The Political Dynamics of Sustainable Coffee', *Journal of Management Studies*, LIII/3 (May 2016), p. 375.

在 1994 年的种族大屠杀中毁于一旦。2002 年，卢旺达出台了一项国家咖啡战略，投资建设水洗站，部分资金来自外国援助计划。烘焙商与农民以及加工商有了直接联系，进而投资培训以及建造检验咖啡品质的杯测实验室。"公平贸易"协会已对经营水洗站的合作社进行了认证。

2006 至 2012 年，卢旺达的咖啡出口总额几乎翻了一番，因为全水洗的咖啡豆的价格高得多。这些额外收入大都流回了生产商的钱包；通过在水洗站的接触，不同种族的农民

一位咖啡杯测师正在卢旺达进行评判。2002 年，卢旺达出台了国家咖啡战略，旨在将卢旺达从商品咖啡生产国转化为精品咖啡生产国。到 2014 年，该国超过 42% 的咖啡豆都经过了全水洗。

之间的紧张关系得到了缓和。[①]如今，卢旺达被公认为精品咖啡生产国；2008年，它成为首个举办国际咖啡杯测赛的非洲国家。

走向全球化的消费

精品咖啡运动对全球咖啡结构带来的最大影响，很可能是促进了新兴经济体消费咖啡，特别是咖啡生产国。咖啡被重新定位成一种令人梦寐以求的饮品，已在追赶西方潮流的年轻消费者中流行起来。

1996年，V. G. 西达尔塔在班加罗尔开设了印度咖啡连锁品牌"每日咖啡"的第一批门店，专为某类目标群体量身打造：

> 我们把这里设计成学生和年轻人出门聚会之地。那时，互联网刚刚兴起。我们认为咖啡馆会吸引班加罗尔具有某种国际视野的软件从业人群。于是，我们开始免费提供上网服务以及供应咖啡。换句话说，你买一杯咖啡，就可以享受30分钟的免费上网服务……70%的印度人口年龄都在35岁以下，而我们80%的顾客年龄也同样如此……他们

---

① Karol C. Boudreaux, 'A Better Brew for Success: Economic Liberalization in Rwanda's Coffee Sector', in *Yes Africa Can: Success Stories from a Dynamic Continent* (World Bank, 2010), pp. 185–99.

希望自己能够获得在电影或电视中看到的，或从互联网了解到的同等体验。[1]

2015年，每日咖啡在印度经营的门店数量超过1500家。其商业计划和顾客诉求的重要内容便是种植自家咖啡树，控制咖啡从作物到成杯的整个流程。

就规模和当前的欠发达状态而言，中国是亚洲最具潜力的咖啡市场。2004至2013年，在中国虽然咖啡的消费量每年增长16%，但人均消费量仍只有83克——在依然以饮茶为主的中国文化里，这个量相当于一年才喝五到六杯咖啡。

而在规模小得多的家庭外部市场，咖啡仅占销售总额的44%。如今，中国已是全球除美国以外拥有星巴克门店数量最多的国家，即便其顾客主要局限于富裕的城市中产阶级。中国消费者尤其喜欢用手机点咖啡作为互赠的"社交礼物"。

目前，中国家庭消费的咖啡几乎全都是以罗布斯塔咖啡为主的速溶产品，其中一半进口自越南。中国的规划发展将带动雀巢、星巴克等大型咖啡企业与云南种植阿拉比卡咖啡树的中国农民合作，预计向这个不断增长的市场提供本地的咖啡产品。

巴西便是生产国自身开发出消费市场的典范。大约95％的成年人都在饮用咖啡，该国几乎要超过美国成为咖

---

① Ashis Mishra, 'Business Model for Indian Retail Sector', *11MB Management Review*, XXV (2013), pp. 165–6.

印度连锁咖啡馆每日咖啡的一家门店，2006 年。

啡消费最大的市场。而巴西国内销售的所有咖啡都必须是由本国种植的。

从 20 世纪 90 年代至 21 世纪 10 年代，咖啡年平均消费率翻了一番。其中很大的动力源于新中产阶级的崛起，他们几乎占了人口的半壁江山。他们欢迎供应卡布奇诺的新式咖啡馆的出现，因此在 2003 至 2009 年，家庭外部市场的消费量增长了 170%。独立认证烘焙与研磨咖啡品质标签的引入使国内咖啡消费发生转变，咖啡数量与品质得到了提升。[1]

印度尼西亚、菲律宾和越南涌现出一批又一批采用本地咖啡豆的烘焙商、零售商以及咖啡连锁店，这些国家是

---

[1] International Coffee Organization, *A Step-by-step Guide to Promote Coffee Consumption in Producing Countries* (London, 2004), pp. 154–207; 'Brazil', www. thecoffeeguide.org, March 2011.

亚洲目前增长最快的咖啡消费市场。越南的中原咖啡公司共经营着五家加工厂，生产速溶咖啡产品，出口60多个国家。与此同时，它在越南拥有或供应1000多家咖啡馆。大多数店都会提供用罗布斯塔咖啡粉和炼乳冲泡的饮品，这对当地顾客群来说具有一定的吸引力。

亚洲最初的几个新兴经济体在咖啡口味上都有所变化。新加坡的"第三波"咖啡文化发展良好，反映了这座城市的全球地位。韩国是当下商业浓缩咖啡机销量增长最快的市场：据说首尔的人均咖啡馆数量比西雅图还多。顾客晚上会待在咖啡馆，逃避回到局促的公寓中，待在店里的平均时间在一小时以上，而咖啡馆的营业时间通常从上午10点到晚上11点。

中国咖啡种植中心云南的一个咖啡树苗圃，2014年。

咖啡新纪元？

在发达国家，精品咖啡运动朝着咖啡商品化的反方向发展，形成抗衡之势。2014 年，美国人均消费量恢复到近4.5 千克（约合 10 磅），从这个回转也可以看出这场咖啡运动的成功。

尽管咖啡行业仍继续由跨国企业主导，但在行业构成及特征上有了重大转变。雀巢公司依然是全球最大的烘焙商，但它最有活力的品牌奈斯派索被定位成了高端精品咖啡品牌。JAB，位于卢森堡的私募股权公司，旗下咖啡品牌组合包括 JDE（Jacobs Douwe Egberts），也在投资精品咖啡，相继收购了皮爷咖啡以及"第三波"咖啡连锁店——知识分子、树墩城和克里格。

如今，全球咖啡消费分布更为均匀，打破了生产国与消费国之间的二元分化。欧洲虽仍是大洲里最大的咖啡市场，约占全球市场的三分之一，但亚洲、北美洲和拉丁美洲当前各占五分之一左右。亚洲的人均消费率大约只有北美洲的十分之一，因此其进一步发展的潜力值得期待。

精品咖啡革命带来的巨大影响，为咖啡新纪元的到来奠定了基础。2014 至 2016 年，世界咖啡年消费量超过了咖啡豆的年产量。从中长期来看，该趋势可能会一直持续，因为新市场的扩张调动了人们对咖啡的需求，而产量却因经济和环境因素下降。

经济发展不仅刺激了咖啡消费，还影响到了咖啡生产。

世界范围的咖啡种植者都在逐渐老去，而他们的子女为了寻找更好的发展机会，又移居到了城市。这一方面可能导致产量下降，但另一方面又可以解决家庭成员在分割小块咖啡农田时闹得不可开交的问题。

2012—2016 年世界咖啡消费的区域分布[1]

地区	占世界总量的比例（%）	复合年增长率（%）
欧洲	33.3	1.2
亚洲及大洋洲	20.9	4.5
北美洲	18.4	2.5
南美洲	16.6	0.4
非洲	7.0	0.9
中美洲	3.5	0.7
世界	100	1.9

气候变化是咖啡种植面临的最大威胁。据估计，到2050 年，全球适合咖啡生产的土地面积将减少一半。[2]除了超极端气温，不断增强的气候波动性也会通过降雨模式、病害与虫害变化影响产量。虽然气候变化可能会导致出现新的咖啡种植区——加州东南部已开始种植咖啡树——但对传统种植区及以此为生的农民将产生深远影响。科学地、有计划地培育更多能够抵抗气候变化及保持良好风味口感的品种，可能有助于减少气候变化所带来的负面影响，但许多农民仍旧需要迁移到别处或转而种植其他作物。

咖啡市场的根基所产生的变化可能会潜在地导致咖啡

---

① 数据源自国际咖啡组织。

② World Coffee Research, *The Future of Coffee: Annual Report 2016*, www.worldcoffeeresearch.org.

价格上涨，尤其对那些能够进入精品咖啡市场的生产商而言。这些生产商已经改变了咖啡行业的地理分布结构。而对消费者来说，精品咖啡革命所带来的积极影响是，他们可以享受到比以往任何时候都更优良的品质以及更多的种类。开开心心泡杯咖啡吧！

埃塞俄比亚商品交易所（ECX）正在进行咖啡交易，2017年。埃塞俄比亚商品交易所旨在让埃塞俄比亚有更好的机会实现其咖啡的内在价值，尽管其监管机制的结构一直存在争议。

食

谱

若认为准备一杯好咖啡一定需要大量高科技设备、相关知识以及塞得鼓鼓囊囊的钱包，则很容易对冲泡咖啡却步。下面提供一些小建议，可在家提高你的咖啡制作水平，无需耗费太多时间或金钱。

保证新鲜度是关键。要一改之前的习惯，开始购买全豆，在制作前才将其研磨。理想的做法是使用可以校准的电动磨盘式磨豆机，研磨出同等大小的颗粒。哪怕是便宜的单"刀片"式磨豆机，也能发挥很大的作用。

少量购买咖啡豆，切勿拖到变质才食用。检查包装袋上的咖啡豆烘焙日期（而非"最佳食用日期"）——通常是在三周内。咖啡豆要在阴凉干燥处以室温储存——切勿放进冰箱。

可以在超市购买咖啡豆，但要在网上搜一搜"精品咖啡（所在的城市）"，找找合适的精品咖啡供应商，或者网上注册订阅服务。购买带有某国标志的"单一产地咖啡豆"，最好选择指定的地区，如埃塞俄比亚的耶加雪菲。

条件允许的话，使用电子秤来量咖啡和水的分量。通常咖啡匙舀 10 克左右的咖啡粉。水温应在 90°C 至 95°C。一个小窍门：将水烧开，然后在倒开水前等待 30 秒。

建议咖啡与水的冲泡比例（近似值——取决于口味）

冲泡方式	饮用人数	咖啡粉（克）	水（毫升）	冲泡时间（分钟）
法压壶	1	20	300	4
60度锥形手冲滤杯	1	18	250	3.5[①]
凯梅克斯手冲咖啡壶	2~3	30	500	4
爱乐压（短款）	1	15	150	0.5
爱乐压（长款）	1	18	250	3.5
意式浓缩咖啡机（单杯）	1~2	7~9	25	25（秒）

法压壶

这是最简单、最好操作的咖啡冲泡方式：粗磨咖啡（类似面包屑粗细），放入壶中，加热水，盖上盖子，放置四分钟（使用计时器），然后按下柱塞，一杯香浓美味的咖啡便做好了——非常适合早餐享用。

滤杯冲泡壶（60度锥形手冲滤杯，凯梅克斯手冲咖啡壶）

若想喝到风味纯净的午后咖啡，不妨试试滤杯冲泡壶，如60度锥形手冲滤杯或凯梅克斯手冲咖啡壶。打湿滤杯，加入中度研磨（类似食盐粗细）的咖啡，水平晃动，注入少量热水浸湿咖啡粉，等待它膨胀或"绽放"。然后每隔一段时间继续轻柔地倒入剩余热水，使之穿过咖啡粉，缓缓滴落。60度锥形手冲滤杯的冲泡时间预计为两分半钟，凯梅克斯手冲咖啡壶则是四分钟。

---

① 此处疑原文有误，与后文提到的冲泡时间矛盾，经查证，应为2.5（即两分半钟）。——译注

爱乐压

爱乐压是极佳的便携式咖啡冲泡设备。润湿滤纸，将滤杯放置滤管上，再置于杯上，倒入中细研磨的咖啡粉，注入热水，搅拌十秒钟，插入活塞压筒，匀速将水压过咖啡粉。若要冲泡一杯浓缩咖啡，接近美式咖啡那种，则用爱乐压的量勺（约15克）加入咖啡粉，将水加到标示的第二级。这种设备介于法压壶和滤杯冲泡壶之间，将活塞压筒插入空管顶部，反转倒置，加入咖啡粉和水，搅拌，浸泡三分钟，将杯子放在顶部，再翻转过来，让水慢慢压过咖啡粉。取出滤杯，用活塞压筒将剩余的咖啡废渣吹出。

摩卡壶（制作炉灶式浓缩咖啡）

炉灶式摩卡壶的窍门在于使下壶刚好置于气阀下方，将细磨咖啡粉倒入滤杯，不求精确。一听到蒸汽通过咖啡粉，萃取到上壶时的啪啪的声音，立马关火。

意式浓缩咖啡机

若不在咖啡机和研磨机上大量投入，就别指望可以做出好的浓缩咖啡。研磨（咖啡粉至类似沙子粗细）至关重要，因为它决定咖啡的流动：将预先研磨好的浓缩拼配咖啡粉置于摩卡壶中，冲泡风味更佳。奶泡的基本原理是将蒸汽管的尖端刚好插入牛奶表面之下，漩涡式搅动，然后慢慢放低奶壶"拉出"奶泡，蒸汽管尖端要始终刚好置于牛奶表面之下。咖啡迷们喜欢摆弄各种机器，而普通消费者去当地咖啡馆可能更方便。

咖啡馆菜单

　　国际咖啡馆菜单主要供应以浓缩咖啡为基础的饮品，搭配蒸牛奶或奶泡。奶泡的种类涵盖了从"干"卡布奇诺中高度透气的大奶泡到馥芮白中带有小气泡的丝绒般微奶泡。每个国家、每间咖啡连锁店的供应分量都各不相同。

美式咖啡　热水加浓缩咖啡

宝贝奇诺　专为儿童准备的不含咖啡的奶泡饮料

拿铁咖啡　浓缩咖啡加蒸牛奶和小杯奶泡。加入糖浆，呈现姜饼、南瓜或香草等风味

卡布奇诺　浓缩咖啡加等份的蒸牛奶和奶泡。咖啡表面撒可可或肉桂粉

冷萃咖啡　冷藏条件下将冷水与咖啡渣浸泡16至24小时后的夏季长饮

冰滴咖啡　以非常缓慢的速度将冷水在咖啡粉中过滤后的长饮——通常为8小时

可塔朵　西班牙式浓缩咖啡（比意式浓缩萃取时间更长，咖啡味道更淡），会加入等份蒸牛奶

意式浓缩咖啡　浓缩咖啡，25~30毫升，在大约9巴的压力下冲泡。如今许多咖啡馆都以双倍浓缩咖啡为标准

馥芮白　"第三波"咖啡浪潮中最受欢迎的牛奶咖啡饮品，

起源于澳大拉西亚[1]。将微奶泡加入双份芮斯崔朵中，表面的"馥芮"通常指最后完成的拿铁拉花

冰咖啡　普通冲泡咖啡冷藏后加冰

玛奇朵　浓缩咖啡加少量奶泡

摩　卡　种类繁多，但基本由浓缩咖啡、巧克力和蒸牛奶制成。通常会附加棉花糖等类似小食

氮气咖啡　注入氮气的冷萃咖啡，具有奶油口感

短笛咖啡　浓缩咖啡加入等份的微奶泡

芮斯崔朵　精萃浓缩咖啡，约15毫升，常见于意大利南部

　　咖啡长期以来一直是制作食物与鸡尾酒的原料。下面介绍的是一些历史食谱，体现了咖啡用途之广。所有内容都是原始做法及说明。

咖啡蛋糕

摘自比顿夫人的《家庭管理手册》（1861）

0.5 磅（约 200 克）黄油

0.5 磅（约 200 克）红糖

0.25 磅（约 100 克）黄金糖浆

0.5 磅（约 200 克）醋栗

1 磅（约 400 克）葡萄干

---

① 澳大拉西亚（Australasia），一般指大洋洲的一个地区，包括澳大利亚、新西兰及其邻近的太平洋岛屿。——译注

1.5 磅（约 600 克）面粉

1 盎司（约 20 克）发酵粉

2 枚鸡蛋

0.5 盎司（约 10 克）混合肉豆蔻、丁香和肉桂咖啡

少量牛奶

将发酵粉、香料连同面粉一起过筛至碗中；加入糖和黄油，一同搅匀，将中间掏空，倒入糖浆，加入大约 0.25 品脱（约 150 毫升）冷却后的浓咖啡，打入鸡蛋，一起打匀；然后用大木勺拌入其他配料，若不够湿润，再加一点牛奶，最后拌进水果，放置在提前准备好的长方形蛋糕盘中烘烤。烘烤一到两个小时，足以做出一个大约 1.75 磅（约 700 克）的蛋糕。

咖啡冰激凌

摘自 A. 埃斯科菲耶的《现代烹饪指南》（1907）

冰激凌准备事宜

将大约 0.67 磅（约 270 克）糖及 10 个蛋黄放入锅中混合至带状。加入 1 夸脱（约 1100 毫升）煮沸的牛奶，一点一点稀释，中火状态下搅拌，直到这些配料都附着在搅拌的勺子上。要避免煮沸，因为这可能会使蛋奶浆分解。将锅里所有配料过滤到盆里，不时搅拌，直至冷却下来。

准备好冰块……用碎冰将盛着配料的盆围起来，混入氯化钠（海盐或盐冻）和硝石。两种盐都可以大大降低水温，迅

速凝结周围液体……进行冷冻的"冰箱"（即容器）……应是纯锡的……把要冷冻的配料倒入其中，抓住盖子把手，来回摇晃器皿，使之不断搅拌……若器皿通过中心轴连接，则可以通过转动把手来保持搅拌的状态。器皿的搅拌运动使配料啪嗒啪嗒不断滴落在冷冻容器四周，并在上面迅速凝结，凝固住的部分很快又用特殊铲子抹走，直到最后形成光滑均匀的整体。

咖啡调味料

在煮沸的牛奶中加入 2 盎司（约 40 克）新烤的碎咖啡籽，浸泡 20 分钟。或者，用等量的咖啡粉和 0.5 品脱（约 300 毫升）的水，混合后形成高浓度饮品，并倒入 1.5 品脱（约 900 毫升）煮沸的牛奶。

提拉米苏

提拉米苏虽被视作意大利经典甜点，但直到 20 世纪 70 年代才诞生于特雷维索一家餐厅。下面是那家餐厅的原始配方，可在网站 www.tiramesu.it 上面浏览，其中配备的小食有手指饼干[①]。

12 个蛋黄

500 克白砂糖

1 千克马斯卡彭芝士

60 根手指饼干

---

① 手指饼干：意大利语中称 savoiardi，英语中称 ladyfinger 或 sponge finger。——译注

咖啡

可可粉

做好咖啡，倒入碗里，放置一旁冷却。将蛋黄与白砂糖搅拌至黏稠，再加入马斯卡彭芝士，做成软奶油。将30根手指饼干浸入咖啡中，注意不要完全浸泡。把饼干排成一排，放于圆盘中央。将一半的奶油涂在这些手指饼干上。剩余饼干和奶油同样如此，将第二层饼干铺在第一层上，然后抹上奶油。再撒上可可粉，冷藏后即可食用。

咖啡火焰鸡尾酒（Bruleau，又名 Brûlot 或 Café Diabolique）

摘自玛莎·麦卡洛克–威廉斯的
《旧日南方的菜肴与饮品》（1913）

将三勺白兰地，连同半个柑橘皮、一打丁香、一根肉桂、几粒多香果以及六块糖一起放入特制的鸡尾酒碗中，酒碗配有白兰地酒勺。放置几个小时后再提取精油。上桌时，要再为每人多舀一勺白兰地，放两块多米诺糖。然后在鸡尾酒碗下面的托盘里倒入酒精，点燃后来回搅动白兰地，直到它也从下部火焰中燃烧起来。等它燃烧两到三分钟——此时若灯光如期熄灭，酒碗散发出的光会非常迷人。三分钟后，倒入浓浓的、滚烫的、清澈的黑咖啡，每人一小杯，继续搅拌，直到火焰熄灭，方可热气腾腾地享用。这种"烧水式"调酒在品种更加丰富的地区被称为咖啡火焰鸡尾酒，起源于新奥尔良，是克里奥尔式烹饪的完美体现。

爱尔兰咖啡

20世纪40年代，一艘横跨大西洋的水上飞机抵达爱尔兰福因斯机场，机场酒保为了使乘客暖和身子，在咖啡中加入了威士忌。当被问及这是否为巴西咖啡时，他的回答是："不，这是爱尔兰咖啡。"此后，这个食谱在爱尔兰和美国得到了进一步传播。下面是泛美咖啡局在1956年出版的书籍《咖啡的乐趣》中推荐的食谱。

在温热的餐酒杯中，加入两茶匙白糖，同时倒入约三分之二的咖啡。搅拌。再加入大约两汤匙的爱尔兰威士忌，然后表面覆上掼奶油。（若想让奶油浮在咖啡表面，试着将之倒在勺子背面——一旦浮在咖啡上，勿再搅拌。）

浓缩咖啡马提尼

30毫升（1杯）浓缩咖啡

50毫升（2盎司）伏特加

10毫升（2茶匙）糖浆

将所有原料放入装满冰块的鸡尾酒调制器中，至少摇晃10秒钟，再倒入经过冰镇的马提尼酒杯里，杯边饰以三颗咖啡豆。

致

谢

写完这本书就像泡了杯浓缩咖啡：要汇集大量材料，方能最后做出让人满意的饮品。

首先感谢迈克尔·利曼委托我来撰写这本书，并给予我足够的时间充分准备，其次感谢 Reaktion 出版社的工作人员，谢谢他们提供的所有帮助。还要感谢阿莱格拉战略咨询公司（Allegra Strategies）、Comunicaffè 咖啡网站、国际咖啡组织、MUMAC 咖啡学院、精品咖啡协会和世界咖啡研究协会与我慷慨分享各种信息。写作过程中，我不断向咖啡行业的诸位学习，大家对我倾囊相助，至今仍令我感到震惊。我要特别感谢阿戈·拉杰里、安娜·哈梅林、安雅·马尔科、亚瑟·欧内斯托·达尔博芬、芭芭拉·达尔博芬、巴里·基瑟、布丽塔·福尔默、克莱夫·梅顿、科林·史密斯、达尔乔·德·卡米利斯、恩佐·弗兰贾莫雷、恩代尔·图兰、恩里科·马尔托尼、肯尼思·麦卡尔平、肯特·巴基、林赛·艾农、路易吉·莫雷洛、毛里齐奥·朱利、罗伯特·瑟斯顿、肖恩·斯泰曼和亚斯明·西尔弗曼，谢谢他们对书中具体内容提出建议。

本书最终得以问世，也离不开妻子伊丽莎白的支持。很荣幸可以与她一同分享我的这杯"咖啡"。

Coffee

⊖

A Global History

Jonathan Morris

Contents

Introduction

Coffee is a global beverage. It is grown commercially on four continents, and consumed enthusiastically in all seven: Antarctic scientists love their coffee. There is even an Italian espresso machine on the International Space Station. Coffee's journey has taken it from the forests of Ethiopia to the *fincas* of Latin America, from Ottoman coffee houses to 'third wave' cafés, and from the coffee pot to the capsule machine.

This book is the first global history of coffee written by a professional historian. It explains how the world acquired a taste for coffee, yet why coffee tastes so different throughout the world. From the beverage's first appearance among Sufi sects in fifteenth-century Arabia, through to the specialty coffee consumers of twenty-first-century Asia, this book discusses who drank coffee, why and where they drank it, how they prepared it and what it tasted like. It identifies the regions and ways in which coffee was grown, who worked the farms and who owned them, and how the beans were processed, traded and transported. It analyses the businesses behind coffee—the brokers, roasters and machine manufacturers—and dissects the geopolitics behind the structures linking producers to consumers.

Regional Distribution of Global Coffee Production (%) [1]

	Africa and Arabia	Caribbean	Asia	Latin America
1700	100	0	0	0
1830	2	38	28	32
1900–1904	1	4	4	91
1970–74	30	3	6	61
2011–15	9	1	32	58

The distinctions between commodity and specialty coffee, and the ways these determine the transformations that coffee undergoes from seed to cup, are the subject of the opening chapter of the book.

The history of coffee is divided into five eras. Coffee first served as the 'Wine of Islam', cultivated on Yemen's mountain terraces and traded among the Muslim peoples around the shores of the Indian Ocean and the Red Sea. Europeans turned it into a colonial good during the eighteenth century, compelling serfs and slaves to plant it in places as far apart as Java and Jamaica.

Coffee was transformed into an industrial product in the second half of the nineteenth century as the rapid expansion of output in Brazil nurtured the development of a mass consumer market in the United States. After the 1950s, coffee became a global commodity as Africa and

---

[1] Data derived from Benoit Daviron and Stefano Ponte, *The Coffee Paradox* (London, 2005), p. 58; ICO data.

Asia regained a significant share of world trade by planting Robusta, a hardier, but harsher-tasting species, used in cheaper blends and soluble products. A movement to recast coffee as a 'specialty beverage' began as a reaction against commodification at the end of the twentieth century. Its transnational success may result in the fifth era of coffee history.

The coffee trade operates with a multitude of definitions and units of measurement. Historical data sets are rarely comparable, so rather than imposing a false unity, the book presents statistics in the form they were originally produced. Macro comparisons, calculated using multiple sources, are intended to convey the direction and scale of change, and numbers should not be regarded as definitive.

Grab a cup, turn the page, and enjoy your journey through coffee's global history.

Chapter 1
Seed to Cup

Coffee is an everyday drink—whether gulped down first thing at breakfast, during a mid-morning break, as an afternoon pick-me-up or as a digestion aid after dinner. Most coffee drinkers have an instinctive sense of what they consider a good cup of coffee, yet few understand what contributes to producing it. This chapter explores coffee's journey from seed to cup. It shows how choices made during this journey determine if the beans are sold as commodity or specialty products.

Commodity or Specialty?

The major reason consumers lack the knowledge to appreciate their coffee is that the industry obscures its complexity and diversity by turning it into a homogenized commodity. Batches of beans harvested at one time are mixed with those picked at another; outputs from farms with different characteristics are combined; sacks from different regions are exported under the same label; green coffees are bought through an exchange where they are never actually seen, before the beans are roasted and blended with others from different countries to be sold under a brand label communicating generic characteristics: 'Rich', 'Mellow' or 'Roaster's Choice'.

Such strategies allow coffee from one source to be substituted with another. Natural events like drought, frost or disease, or man-made ones such as war, can set back coffee production in a region for years. Farmers, exporters, brokers and roasters use homogenization as a risk-management strategy. At least 90 per cent of world coffee production enters the commodity sector.

The remaining 5–10 per cent is 'specialty coffee': high-quality coffee with a distinctive flavour profile and identifiable geographical origins. Like wine, a coffee's flavour is reflective of the variety grown, the district's micro-environment (*terroir*), the growing season's prevalent climatic conditions, and the care with which it is harvested, processed, stored and shipped. Wine contains around three hundred compounds affecting its flavour; for coffee the figure is estimated to be well over a thousand. This 'special(i)ty' sector (Europeans used to include the 'i', Americans don't) has grown exponentially over the last thirty years.

Coffee farms can be divided into three types. Large agri-businesses located on huge Brazilian estates constitute less than 1 per cent of all farms, but produce roughly 10 per cent of the world's supply. Family-owned enterprises, commonly found in Central America and Colombia, make up 5 per cent of farms but account for 30 per cent of output, much of specialty quality. Smallholdings of fewer than 5 hectares (12 acres) number 95 per cent of all coffee farms and produce 60 per cent of global output. On most of these farms coffee is grown as a cash crop within a subsistence regime. [1]

Arabica

Coffee is a gift from Africa, where over 130 species of the genus *Coffea* have been identified. The Arabica coffee plant, *Coffea arabica*, evolved in the southwestern Ethiopian high-lands and bordering regions of Kenya and South Sudan, where it still grows wild today. Today Arabica is grown commercially throughout the tropics. It cannot survive outside this belt as the plants die if the temperature falls below freezing. Arabica was the first—and until the twentieth century, the only—species of coffee grown for human consumption. Currently it accounts for around two-thirds of world production.

The Arabica coffee plant is a woody perennial evergreen shrub that, in the wild, grows 9–12 metres (30–40 ft) high under the forest's semi-shade canopy. It is commonly inaccurately referred to as a tree. It is self-pollinating, producing a cluster of small, white, fragrant flowers. The number and size of flowers is largely weather dependent.

[1] David Browning and Shirin Moayyad, 'Social Sustainability', in *The Craft and Science of Coffee*, ed. B. Folmer (Amsterdam, 2017), p. 109.

Showers spark the plant's blossom, but the fruits set best in dry conditions. In semi-dry climates, there is one flowering season; where rainfall is greater there may be two or more, with fruit and flowers simultaneously on the plant. The seeds at the flower's base develop into drupes, referred to as coffee cherries. These ripen 30–35 weeks after flowering, changing colour from green to deep red (yellow in some varieties), at which point they are ready to harvest.

Each cherry contains two flat-faced seeds, commonly, if inaccurately, referred to as beans. Each seed is covered by a parchment layer, and protected by sweet, soft, pinkish pulp under the skin. Occasionally a single rounded seed develops. This is known as a peaberry. Producers separate these from the rest of the crop to be sold at a premium, arguing that they offer better sensory qualities and that the shape is conducive to even roasting. Sceptics suggest the higher price compensates for the fact that a peaberry weighs less than two ordinary beans.

Robusta

During the last three decades of the nineteenth century, the coffee world was transformed by a devastating outbreak of leaf rust that virtually wiped out production in Asia. Coffee cultivators, notably in the Dutch East Indies, started searching for alternative species. They tried *Coffea liberica*, or Liberian coffee, but this too proved susceptible to rust. They then shifted to *Coffea canephora*, known as Robusta, which was sourced from the Congo, via Belgium.

Robusta is not only rust resistant, but it tolerates higher temperatures and humidity than Arabica, making it capable of flourishing at lower elevations. The tree has an umbrella shape, with smaller but more numerous cherries gathered in clusters, making it easier to harvest. Its easy cultivation enabled it to be used as an entrée into coffee production, most recently by Vietnam. Currently Robusta forms around 35–40 per cent of world output.

Robusta suffers from one major defect: it produces poorer quality coffee than Arabica. A common Robusta tasting note descriptor is burnt rubber. It is nearly always used as part of a blend, and is often used in instant or soluble coffee products. Coffees marketed as specialty Robustas (notably from India) are usually the result of better husbandry and processing practices. Robusta also contains twice the caffeine levels of Arabica.

Varieties

For most of coffee history, only two varieties of Arabica were cultivated. The commercial variety closest to wild Ethiopian coffee is known as Typica. Bourbon, a natural mutation that occurred in the colony where the French first planted coffee, is the other. Bourbon is higher yielding and tends to produce fruit flavours in contrast to Typica's floral ones.

Brazilian coffee researchers developed the dwarf cultivars Caturra and Catuai, which became popular in the post-war era because of their high yields and easier cultivation. Caturra was interbred with a naturally occurring Arabica-Robusta hybrid, Timor, to produce Catimor, another high-yielding dwarf with greater disease resistance that could be planted at lower elevations. These were popular among commodity farmers but coffee aficionados were sceptical about their quality in the cup.

During the twenty-first century, however, the specialty world has shown much greater interest in varieties. Geisha (or Gesha), a natural Ethiopian variety, is primarily responsible for this. It was introduced into Central America in the 1950s but gained few followers due to its low yields. In 2004, however, the Peterson family, new owners of the La Esmerelda estate in Boquete, Panama, identified these beans as the source of their farm's individual cup flavour. They separated them out and entered them into the Specialty Coffee Association of America's Roaster's Guild cupping competition, winning three years in a row. In 2007 one lot sold for over one hundred times the price of commodity coffee.

Unsurprisingly Geisha, which produces a cup of complex aromatic and floral flavours with a body like tea, is now being grown on many farms, while other varieties such as Pacamara and Yellow Bourbon have become sought after. In 2018 a natural processed Geisha from Panama was auctioned for a new world record price of $803/lb, when the commodity coffee price was $1.11/lb. [1]

Terroir

The micro-environment, or *terroir*, where a coffee crop is cultivated has a substantial effect on its flavour profile. Discerning a direct line between any one single factor and outcomes in the cup is hazardous, due to the difficulty of isolating the multiple variables involved.

The most important elements are temperature and elevation. Arabica becomes unproductive at temperatures over 32°c, so elevation is essential to avoid these. In the equatorial zone within 10 degrees latitude of the Equator, suitable conditions are usually found above 900 metres (2,950 ft); at the subtropical extremes, elevations can be much lower. The best Ethiopian coffees, such as those from the Yirgacheffe district, are grown at around 1,800 metres (5,905 ft), whereas Hawaii's celebrated Kona coffee belt starts 200 metres (656 ft) above sea level.

Within a region, the higher the altitude at which coffee is grown, the better the bean quality. Higher-grown coffees possess more concentrated flavours, possibly due to the greater difference between day-and night-time temperatures. Beans grown at lower levels are softer, less dense and age more rapidly. El Salvador grades its coffee by elevation; the top-grade

[1] Nick Brown, 'Natural Geisha Breaks Best of Panama Auction Record at $803 per pound', www.dailycoffeenews.com, 20 July 2018.

'Strictly High Grown' is grown above 1,200 metres (3,937 ft).

The average air temperature has a significant effect on the sensory profile of the bean, with cooler temperatures being correlated with positive features such as acidity (experienced on the palette as liveliness) and fruit flavours, whereas high temperatures can result in reduced aromatic quality and an increase in off-flavours. Surveys of the main Arabica-producing areas suggest that the ideal conditions for producing specialty coffee are found in zones where the temperature is relatively constant throughout the year, ranging from 13°c in the coolest month to 25°c in the hottest. These conditions obtain only in a quarter of the land used for Arabica cultivation. [1]

Coffee can be grown in a wide variety of soil types, providing they are deep, well drained and rich in nutrients. Volcanic soils are particularly favoured and some believe these correlate with greater acidity. Similarly, coffee flourishes in a wide variety of climate regimes, providing it receives the equivalent of 125 centimetres (49 in.) of rainfall a year. This can be distributed either evenly or seasonally, or delivered artificially via irrigation systems.

Coffee grows naturally in partial shade, requiring only 25 per cent sunlight to develop efficiently. The impact of shade on taste profiles is contentious. Research in Costa Rica suggested shade promoted better acidity, less bitterness and lower astringency, yet a study in Colombia suggested

---

[1] Charles Lambot et al., 'Cultivating Coffee Quality', in *Craft and Science*, ed. Folmer, pp. 21–2.

the reverse, while one in Hawaii found no difference in cup quality between shade and sun-grown coffee of the same type. [1]

Cultivation

Farmers cultivate coffee from seed, and transplant potted plants into fields at around eighteen months. Saplings usually begin fruiting at three to four years of age and reach commercial maturity at around five to seven years. The trees develop vertical shoots from which lateral spurs grow outwards. Fruit sets in clusters along the previous year's laterals. Pruning restricts the trees' height to 8–10 metres (26–33 ft) to facilitate picking.

There is no theoretical limit to the productive life of a tree providing it remains healthy. However, when stressed (by lack of food or water), the coffee plant will literally sacrifice itself to preserve that year's crop, letting its leaves yellow and its branches die back to the point of no recovery.

Shade alleviates stress, moderating air and soil temperature extremes and reducing the plants' need for food. In the absence of natural shade, trees can be planted to serve as wind-breaks and stabilize slopes, preventing soil erosion. Peasant farmers frequently shade their coffee plants with subsistence crops.

---

[1] Shawn Steiman, 'Why Does Coffee Taste That Way', in *Coffee: A Comprehensive Guide*, ed. R. Thurston, J. Morris and S. Steiman (Lanham, MD, 2013), P. 298.

Growers wanting high yields often adopt dense planting schemes where coffee trees self-shade each other, creating 'coffee hedges'. This drives down individual plant yield, but significantly increases the overall yield per hectare. This approach lends itself to mechanized farming and is commonly found on large Brazilian commercial coffee plantations. Such 'sun-grown' or 'technified' coffee requires greater amounts of fertilizers and frequent weeding and is more susceptible to disease, not least due to the reduction in birdlife preying on insects.

Coffee leaf rust caused by the fungus *Hemileia vastatrix* is the most damaging disease prevalent in coffee. Orange and yellow spots appear on the leaves which then drop off. Defoliation kills branches and eventually the plant itself. An outbreak in the Central Americas in 2011 affected 70 per cent of farms over five years, causing 1.7 million coffee workers to lose their jobs. [1] The coffee berry borer, a small black beetle that lays its eggs at the coffee cherry's centre, is the chief insect threat. Once the larvae hatch they eat their way out of the bean. An acute infestation can destroy around 50 per cent of the crop.

Harvesting

While *terroir* and variety are responsible for the coffee's

---

[1] 'Applied R and D for Coffee Leaf Rust', www.worldcoffeeresearch.org, accessed 10 December 2004.

flavour profile, it is primarily harvesting and processing practices that determine the quality of the lot produced.

The coffee cherries are either picked or stripped from the tree. A high-quality coffee can only be produced from a consistent batch of fully ripened beans, so pickers select individual cherries, leaving others on the branch to reach maturity. Selective picking is labour and cost intensive. It requires small producers to band together, helping harvest each other's crops, while larger estates recruit seasonal workers, and incentivize them to exercise quality control while picking.

Stripping, by contrast, involves removing all the berries from the branch by grasping it with one hand while running the other hand down its length. The cherries (and other debris) land on the ground, or in pre-positioned nets, and are then sorted for processing. Commodity market producers generally use the stripping method, taking advantage of the relative pre-dictability of rainy (hence flowering) periods, to estimate when around 75 per cent of the crop will have matured.

On flatland plantations, harvesting machines perform a mechanical form of stripping, passing along the coffee hedges. It has been calculated that five workers using machinery on a Brazilian plantation can harvest the same amount of coffee in three days as a thousand pickers working for one day in the mountainous regions of Guatemala. [1]

---

[1] Oxfam, *Mugged: Poverty in Your Coffee Cup* (Oxford, 2002), p. 20.

Natural Processing

Processing removes the cherry's protective layers surrounding the bean. First the skin and pulp surrounding the stone are removed using either a 'dry' or 'wet' processing method. The coffee is then sent for milling to remove the remaining parchment left covering the bean.

Natural, or 'dry', processing involves spreading the coffee cherries out on a concrete patio in the sun, and periodically turning them as the fruit dries out and decomposes. Using rakes, the coffee is shaped into rows of about 5 centimetres (2 in.) in height, with patches of bare patio between them. The coffee is periodically shifted into the bare space, leaving the space previously occupied by the coffee to dry. The whole process takes around two weeks, after which the beans are ready for hulling.

Natural processing is particularly suited to environments where water is in short supply. In the Yemen, where the technique first developed, coffee can be seen drying on the flat roofs of the houses in mountain villages where it is grown. The process accentuates the body and fruitiness of the coffee, as well as creating a 'wild' tang in the aftertaste. Done well, the flavours can be exhilarating; done badly, they can recall a farmyard.

The main attraction of natural processing is its cost-effectiveness. Most commodity coffee is dry-processed:

both Brazilian 'natural' Arabica, and nearly all the world's Robusta. From the quality perspective, the difficulty with the dry process is lack of consistency. The harvested cherries are usually only sorted manually beforehand, increasing the danger that bad cherries will get through and contaminate the rest of the batch. The process is also uneven, with fruits experiencing a variety of temperatures, and there are dangers of excessive fermentation or mould.

Washed

Wet processing tends to produce a smoother coffee with more consistent taste and better acidity. The process begins with the cherries being placed in a flotation tank. The dense ripe cherries sink to the trough's bottom, while both overripe and underdeveloped cherries float on top, along with any detritus like sticks and leaves. These 'floaters' are removed and a pumping system conveys the submerged cherries into a mechanical de-pulper. Here they are broken against a screen, which prevents the skin and pith passing through with the bean. Underripe cherries are too tough to be broken by the screen, and are removed at this point.

Beans are sent to tanks of clean water where they sit for twelve to fourteen hours while the bean's sticky mucilage is broken down by fermentation. Assessment is often carried out by feeling a handful of beans: when the mucilage has disappeared, the beans are removed.

Wet processing can involve considerable amounts of water, and in recent years both de-pulping and de-mucilator machines have been developed to reduce this. De-mucilator machines sit directly behind the de-pulping screen, rubbing the beans against each other to remove the mucilage, avoiding the need for fermentation.

Once the mucilage is removed, the beans are rinsed in water and then dried down to 11 per cent moisture content. This will be done on either patios or tables, sometimes situated in clear plastic tunnels to harness the greenhouse effect while protecting the coffee. Mechanical dryers are used where the climate dictates.

Pulped Natural and Honey Coffee

The pulped natural method developed in Brazil in the 1980s. In this process the beans were passed through the de-pulper, but then sent directly to the drying patio with their mucilage still attached. It resulted in some outstanding coffees, combining the body obtained from dry processing with the acidity characteristic of wet. Central American producers adopted and modified the process to create honey coffees, which are dried slowly in a more humid environment, developing their aromatic qualities. These are divided into yellow, red and black categories, depending on the amount of mucilage left on the bean—the black retaining all its mucilage and taking up to thirty days to dry.

Animal Processing

Indonesian coffee known as Kopi Luwak is famously processed by palm civet cats. They eat cherries that have fallen to the ground and then excrete beans, effectively performing the de-pulping process. These are collected, washed and finished in the usual way. The enzymes in the cat's digestive system supposedly impart a unique flavour, enabling the coffee to be retailed for upwards of $100/lb, while a single cup of Kopi Luwak in New York can cost $30. Unsurprisingly other countries have now discovered animals capable of performing the same function, including a Vietnamese weasel, Thai elephants and the Brazilian Jacu bird.

Sadly, the fad for Kopi Luwak has resulted in many of the animals being captured and caged to be force-fed coffee beans. Wild certifications do exist but a large proportion of what is sold is presented under false pretences, sometimes having been chemically treated. The bottom line with animal-processed coffee is that what comes out reflects what goes in, and producers can cut costs by using cheaper cherries. The usual qualities ascribed to Kopi Luwak are low acidity and lack of bitterness—these are more about moderating bad coffee than enhancing the good stuff.

Resting

After the beans have been processed and dried, they

will ideally be left to 'rest' in their parchment, stored away from the elements, for one to two months prior to milling. This allows them to mature, losing the grassy taste of freshly processed coffee.

In India's southwestern Malabar region, resting coffee is stored in open warehouses exposed to monsoon winds and humidity. This results in the beans changing colour to a golden yellow and swelling up as their moisture content rises to 13–14 per cent. The results mimic the effects of the long sea journeys undertaken by coffee during the nineteenth century, and produce a mellow coffee low in acidity.

Milling

After resting, the remaining parchment is removed from the bean by a hulling machine. The coffee is then graded by size, using a tilted sorting table. Standard sizes are measured in diameters of 64ths of an inch (usually 10–20). Kenyan AA is size 18, for example. Defective beans are also removed during milling, either by hand or using a colour-sorting machine to identify unripe, broken or insect-damaged beans.

After milling is completed, the green coffee is packed into burlap or hessian sacks made from jute. These protect the coffee from sun and rain, while allowing air to flow through the beans, thus avoiding mildew. The air circulation can give rise to staling, and petroleum-based coatings can leave a so-called 'baggy' taste. In recent years, specialty producers have started placing green coffee into large multilayered plastic bags that create a gas-and moisture-proof barrier before sealing them into the sacks. The standard sack size is 60 kilograms (132 lb). Many trade statistics are quantified in sacks, rather than by weight.

Classification

By the time the bags arrive at the shipping port, the coffee will have changed hands several times. Washing stations and milling plants require capital investment and production volumes that small farmers cannot afford. In some origins, nearly all the supply chain links are in private hands, with growers selling their beans to middlemen at the farm gate; in others, farmers operate cooperative washing stations, and sell their coffee collectively to mills. Export agencies are the last recipients of the processed coffee in the country of origin, preparing and packing it to be shipped to overseas consumer markets.

At each stage along the supply chain, individual lots become mixed into bigger batches based on the standardized categories utilized by the producer country. Brazil, for example, recognizes seven coffee grades defined principally in relation to the number of permitted defects within a sample: these include the number of overripe, broken or diseased beans, as well as stones, husks, twigs and so on. These broad classifications can be related to the prevailing prices operating in world markets, thus allowing coffee to be traded as a commodity.

International Trading

This is done through the two leading coffee futures trading markets: the New York International Commodity Exchange (ICE) for Arabica and the London International Financial Futures and Options Exchange (LIFFE) for Robusta. In both, a standard contract is traded for the delivery of mainstream coffee at a predetermined future date. In New York the traded contract is the 'C' contract: a consignment delivery of 37,500 lbs (or 17,000 kg) of washed mild Arabica coffee with a rating of 9–23 defects within a 12 oz (or 350 g) sample. Lots with fewer defects are traded at a premium, while below-par lots are sold at a discount. The 'C' contract covers coffee from twenty producer countries, with some, such as Colombia, enjoying automatic premiums, while others, including the Dominican Republic, are discounted. Ten delivery months are quoted over a two-year period, with a trading position closed a month before delivery.

These futures exchanges are important for the coffee-trading industry. As the exchange logs all contracts agreed between brokers for future coffee deliveries, an indicator price emerges. Some, though few, of these contracts result in physical coffee deliveries as per the exchange's rules and standard contracts. A greater volume of physical coffee is traded via forward contracts, incorporating the standard futures price as a base point from which the final price for a particular grade of coffee is calculated. So a tender including a differential might be quoted for delivery of grade x from country y for shipment in October at New York December +10 (that is, the price for the December delivered 'C' contract' plus 10 points).

This leaves uncertainty over the final price, which can be offset or hedged by using the exchange's futures and options markets. Trading in options (the option, not the obligation, to make or take delivery of a futures contract at a certain price threshold or expiration period) began in 1986 in New York. It has brought many speculators into the market, to the point that in 2015 the volume of contracts traded was 27 times that of world Arabica production.[1] This gap between the physical and traded volumes of green coffee is beneficial to traders, as the liquidity generated means it is easier for brokers to hedge their positions. They transfer price fluctuation risks from themselves to financial speculators. Conversely, increasing speculation results in greater price volatility, which impacts others in the value chain, such as growers and small roasters, who do not have access to the futures market.

[1] Eric Nadelberg et al., 'Trading and Transaction', in *Craft and Science*, ed. Folmer, p. 207.

Shipping

Brokers acquire control of green coffee by buying in spot markets at either exporting or importing ports. The commodity coffee market is dominated by a few leading brokerage networks. The largest is the Hamburg-based Neumann Kaffee Group, which handles 10 per cent of world coffee demand. Brokerages manage coffee movement logistics, such as warehousing and shipping, and delivering to roasters who have usually purchased via contract options. Coffee is shipped in containers, each one holding around 275 bags of coffee. Entire shipments destined for one roaster may be loaded into a single container, with the beans air blown into it and sucked out into silos at the journey's end.

The major U.S. receiving ports are New York and New Orleans. (The latter's closure in 2005 due to Hurricane Katrina provoked a brief scare that America would run out of coffee.) In Europe, the leading centre is Antwerp, which warehouses roughly half the coffee arriving into the Continent.

Decaffeination

Decaffeination plants are usually located in port hinterlands. Green beans are steamed then soaked in hot water, inducing them to swell up. This enables the solvents used—methylene chloride or ethyl acetate—to remove the caffeine from the beans. The solvent is drained, the caffeine extract sold off and the beans steam cleaned and dried. Liquid or supercritical carbon dioxide can also be used under very high pressures. This is costly but removes fewer additional compounds. Alternatively, using a process developed in Switzerland, hot water can be used as the solvent. After eight hours or so, the beans are drained, and the water is passed through charcoal filter beds to remove the caffeine. The water is then concentrated, and returned to the beans so they can reabsorb the remaining flavours. The process is performed using batches of beans with the water from one lot being returned to a subsequent one.

Roasting

The retail coffee-roasting business is dominated by large, multinational food conglomerates such as Nestlé, Jacobs Douwe Egberts, J. M. Smucker (owners of Folgers), Kraft Heinz (Maxwell House) and Tchibo. Medium-sized roasters frequently produce their own brand labelled coffee for supermarket retailers and grocery chains, and supply the horeca (hotel, restaurant and catering) trade. Niche roasters often concentrate on specialty coffees, including single origins and estates, while micro roasters produce small batches of coffee primarily for sale in their own shops and local supply networks.

The basic principle of coffee roasting is that beans are

heated evenly to a final temperature of between 200°c and 250°c. Most roasters use a so-called drum roaster in which the beans are rotated in a metal cylinder situated above a flame. These machines usually roast one batch at a time, with each batch lasting between eight and twenty minutes, depending on the finish required. Large industrial producers may use continuously operating fluid bed roasters in which the beans are blasted with pressurized jets of hot air for around two minutes.

During roasting the green beans turn yellow, then light brown as they lose their moisture and the starches are caramelized and converted to sugars. A 'first crack' is heard at around 205°c as the beans break open under pressure from the gases building up inside them. The beans continue to brown until a second crack occurs at around 225°c as the cell walls collapse and glossy oils are exuded onto the bean's surface. If roasting is continued, the remaining sugars will carbonize, and the coffee turns black.

The two main factors in determining the coffee's flavour are the degree of roast and the speed of roasting. Artisan roasters will adjust their machines according to their interpretations of the sounds and smells emanating from their machine, as well as the appearance of beans extracted using a sampler. Many machines can be programmed to perform a particular roast to order.

Lighter roasts, developed slowly to a point somewhere between the first and second crack, are believed to bring out the best in most specialty coffees. Aromas, flavours and acidity will decrease with further roasting, whereas body and bitterness increase. Sweetness peaks somewhere between the first and second crack as the sugars caramelize before starting to turn bitter. After the second crack, the flavours from the roast overpower those from the beans, which is why dark roasts are often used for lesser-quality coffees such as Robusta.

Once the roast is completed, the beans must be cooled as quickly as possible to prevent further cooking: most drum roasting machines deposit them into a perforated tray where they are stirred while cool air is drawn through them. Alternatively, the beans may be sprayed or dipped in water. As the beans absorb water, commodity roasters will often use cooling as a pretext for increasing the weight (and hence value) of the coffee.

Instant

Instant or soluble coffee is produced by freeze-drying or spray drying. An extract is prepared from roast and ground coffee with pressurised water at 175°c, concentrated, and frozen to minus 40°c. It is broken up into small granules, and dried in a heated vacuum chamber. The frozen water within the granules sublimates directly into vapour that is removed by condenser. Spray drying, the older and cheaper process, requires the concentrate to be sprayed into a drying tower, where it turns into a dehydrated powder as it falls through a stream of air heated to 250°c.

Blending

Roasters blend to develop distinctive taste profiles for their brands and to manage costs. Coffee blends are usually based on unwashed Brazilian Arabicas (known as Naturals or Santos) whose relative neutrality means they can form the blend's bulk. Character is then added by incorporating Colombian or other so-called Milds with more distinctive features. Low-cost, mass-market blends begin with a Robusta base, adding Santos and topping out with coffee from other origins. High-quality blends usually combine the beans after each lot has been roasted to its own sweet point; however, it is more cost-efficient to combine the beans before and roast them together, so mass-market roasters use a one-size-fits-all approach.

The roaster's art lies in adjusting the roasts and blends of the available coffee so that the taste profile for a brand remains consistent even if its contents have changed. Often this involves substituting one coffee type with another: some roasters agree forward contracts with brokers in which the coffee delivered can be varied according to availability. For example, a Robusta contract might allow for either Ugandan standard grade or Ivory Coast grade 2.

Cupping

After roasting, the coffee's flavour notes acquired throughout its cultivation and processing can be appreciated. Roasters check for quality and consistency by 'cupping' the final product. The same procedure is used to evaluate samples from brokers, and by buyers and producers in the origin country. Comparative cupping uses identically sized parcels of green coffees in a sample roaster so each is prepared similarly. The beans are ground and placed into porcelain cups. Cuppers evaluate the aroma of the dry samples, after which hot water is added. The cuppers then check the wet aroma. After a four-minute wait, they use a spoon to 'break' the surface's crust, putting their noses as close as possible to the cup to catch the aroma, and then slurp spoonfuls of coffee with as much air as possible to spray the coffee around the tongue's taste buds.

The Coffee Quality Institute's Q cupping scheme has been adopted by the specialty coffee world as a method of standardizing cuppers' assessments. All certified Q graders are trained and examined to ensure consistency. Scores are allotted for fragrance/aroma (dry and wet), acidity, flavour, body, aftertaste, uniformity, clean cup, balance, sweetness and overall impression, after which deductions are made for defects to give an overall score out of 100. Coffees scoring over 80 are considered specialty standard.

Packaging

Before it leaves the roastery, the coffee will be allowed to de-gas (expelling the carbon dioxide build-up from roasting) for up to three days before being packaged into airtight or vacuum packs. Most coffee is now pre-ground at the factory for consumer convenience, although this accelerates staling once the pack is opened, because a greater proportion of the coffee's surface area is exposed to oxygen. Higher-grade coffee is often sealed into bags with a one-way valve, allowing the beans to continue expelling gas without air entering, while some coffees are nitrogen-flushed to remove oxygen from the package entirely.

Brewing

There are innumerable procedures and equipment for brewing coffee. They can be divided into four basic methods:

1. Boiling—ground coffee and water are heated together, as with Turkish coffee;

2. Immersion—hot water is added to the coffee, as with French press;

3. Percolation—hot water is passed through the grounds, as with filter;

4. Pressure brewing—hot water is passed with pressure through the grounds, speeding up extraction rates, as with espresso.

An industry saying is that it takes a year for coffee to travel from field to cup, and a minute for the consumer to mess it up. To avoid doing so, see the Recipes section in this book.

Health

The brewing method chosen is one determinant of the caffeine level of the final beverage. The longer water is in contact with the coffee, the more caffeine it absorbs. Other key determinants are beverage size and, most importantly, the proportion of Robusta in the blend, as this contains twice the caffeine of Arabica. The U.S. government's *Dietary Guidelines for Americans, 2015–2020* suggest that a standard 8 fl. oz (235 ml) serving of drip-brewed coffee would contain 96 mg of caffeine, a similar sized cup of instant 66 mg, and a 1 fl. oz (30 ml) single espresso 64 mg. Such average figures vary wildly, but it is worth noting that the guidelines also state that consumption of up to 400 mg a day can be regarded as healthy. [1]

Caffeine is a stimulant that can increase brain function and combat drowsiness. Brain activity is normally regulated by adenosine, which binds to receptors on the surfaces of nerve cells, inducing sleep and causing the blood vessels to dilate to receive more oxygen. When caffeine crosses

---

[1] U.S. Department of Health and Human Services and U.S. Department of Agriculture, *2015–2020 Dietary Guidelines for Americans*, 8th edn (2015), p. 33, www.health.gov/dietaryguidelines/2015/guidelines.

from the bloodstream into the brain, it binds to the nerve cells' receptors, preventing the adenosine from reaching them, while the blood vessels contract, reducing headaches. Adrenaline production is increased, raising alertness, and dopamine reabsorption is slowed down, heightening a sense of pleasure. Therefore, coffee drinkers can use the beverage to 'kick start' the morning, keep them awake on the night shift, relieve stress headaches or simply to feel good.

Consuming too much caffeine over a short period can be detrimental to health. It can increase heart rate and blood pressure, provoking a caffeine-induced version of 'the jitters', accompanied by symptoms like light-headedness, anxiety, insomnia and diarrhoea. The question is, how much is *too* much? Caffeine concentration in the brain reaches its maximum around one hour after drinking coffee, while its usual half-life in the body is three to four hours. Caffeine metabolization varies between individuals, with factors such as weight, genetics, gender and lifestyle contributing significantly. Overconsumption of caffeine can build dependency (hence withdrawal headaches for 'java junkies'), but it can also build tolerance to its effects.

The combination of caffeine's variation in a single serving of coffee and the range of individual responses to the drug explain why it is almost impossible to produce any meaningful recommendations about suitable levels of coffee consumption. Public health surveys are bedevilled by reporting problems. One person's cup of coffee will be a different size, blend and brewing process to another's, even before accounting for any impact of milk and sugar.

Current studies are discovering many positive associations with coffee consumption, suggesting that coffee's chemicals have a role in protecting against liver disease, kidney, bowel and, to a lesser extent, breast cancer, and that coffee may decrease the risk of Alzheimer's and Parkinson's diseases because of its high levels of antioxidants. Coffee has been cleared of having a dehydrating or diuretic effect and may assist in combating adult-onset diabetes. Recent studies suggest that coffee drinkers live longer. According to the editor of the *American Journal of Medicine*, coffee lovers should 'partake and enjoy this mild and perhaps beneficial addiction'. [1]

---

[1] Joseph Alpert, 'Hey Doc, is it OK for me to Drink Coffee?', *American Journal of Medicine*, CXXII/7 (2009), pp. 597–8.

Chapter 2
The Origin of a Coffee Economy

Coffee has a 'foundation myth', much beloved by marketers, that one day Kaldi, a young Ethiopian goatherd, noticed his animals became agitated after eating a shrub's red berries. Kaldi chewed the berries himself and ended up 'dancing' around. Kaldi was then either discovered by, or went to consult with, an imam, who also sampled the berries. He either a) found they kept him awake during late-night prayers so turned them into an infusion to share with others; or b) threw them in the fire in disgust only to smell their delicious aroma, deciding to retrieve them from the embers, grind them up, add hot water and drink the resulting beverage!

The Kaldi story first appeared in Europe in 1671 as part of a coffee treatise published by Antonio Fausto Naironi, a Maronite Christian from the Levant (today's Lebanon) who had emigrated to Rome. He likely heard it in his homeland. Exactly when, where and in what forms humans first came to consume coffee cannot be definitively established. There are rumours of charred beans being found at ancient sites, and some suggest herbs and decoctions described in the *Canon of Medicine* by the Persian physician and philosopher Ibn Sīnā (980–1037), also known as Avicenna, derive from the coffee plant.

It is certain that for the first two hundred or so years of coffee's recorded existence, between 1450 and 1650, it was consumed almost exclusively by Muslim peoples whose custom sustained a coffee economy centred around the Red Sea. This was the world from which modern versions of the drink evolved and the foundations of the contemporary coffee house format laid.

The Oromo tribe, occupying a large swathe of southern Ethiopia, including the Kaffa and Buno regions in which Arabica coffee is indigenous, prepare a variety of foodstuffs and beverages utilizing different elements of the plant. These include *kuti*, tea made from lightly roasted young plant leaves, *hoja*, combining the berry's dried skins with cow's milk, and *bunna qela*, in which dried coffee beans are roasted with butter and salt to produce a solid stimulating snack, carried on expeditions and eaten to heighten energy levels.

Buna is the most well known. Dried coffee husks are simmered in boiling water for fifteen minutes before the resultant beverage is served. Today, coffee farmers have started selling a similar product named *cascara*, consisting of the dried cherry skins removed during processing, brewed as a fruit tea. Originally the beverage was prepared with the cherry's entire desiccated remains—skin, pulp and stone.

Named *qishr* in Arabic, this infusion appears to have made its way across the 32 kilometres (20 mi.) of the Babel-Mandeb straits at the southern end of the Red Sea during the mid-fifteenth century. It was adopted by the Sufi mystic sects in Yemen for use in *dhikrs*. A stimulating potion called *qahwa*, was incorporated into the ritual's beginning, ladled out by the leader from a large vessel and passed around while the group chanted a mantra. *Qahwa* was vital because Sufism was practised by laymen who worked during the day: the etymology of the word implies a lessening of desire, presumably for sleep.

Qahwa was originally prepared with *kafta*, the leaves of the *khat* plant. This has hallucinogenic properties that promote a sense of euphoria, but using *qishr* would have assisted in keeping worshippers awake. The switch was supposedly instigated by the Sufi mufti Muhammed al-Dhabani, who died in 1470. He is the first historical personage we can associate with coffee. The Arab scholar Abd al-Qadir al-Jaziri, whose manuscript *Umdat al safwa fi hill al-qahwa*, written around 1556, is the principal information source on coffee's spread in the Islamic world and reproduces an account claiming that al-Dhabani travelled to Ethiopia, notes that he found the people using *qahwa* though he knew nothing of its characteristics. After he had returned to Aden, he felt ill, and remembering [*qahwa*], he drank it and benefitted by it. He found that among its properties was that it drove away fatigue and lethargy, and brought to the body a certain sprightliness and vigour. In consequence, when he became a Sufi, he and other Sufis in Aden began to use the beverage made from it. [1]

A second account suggests that although Ali ibn 'Umar al-Shadhili is celebrated as the 'father' of *qahwa* in Mocha, this was made from *khat*, whereas in:

> Aden at the time of . . . al-Dhabani there was no *kafta*, so he said to those that followed him . . . that 'coffee beans . . . promoted wakefulness, so try *qahwa* made from it'. They tried it, and found that it performed the same function . . . with little expense or trouble. [2]

Qahwa initially referred simply to the religious potion, but subsequently became the term for Arabic coffee prepared with beans alone, whereas *qishr* still refers to an infusion of dried fruits and spices.

Sufi practices helped transport coffee knowledge northwards into the Arabian territories, the Hijaz, on the eastern shores of the Red Sea. These included the holy cities of Mecca, Jeddah and Medina. Coffee eventually arrived at Cairo, capital of the ruling Mamluk sultanate, sometime during the 1500s, where it was first used by Yemeni students at the Al-Azhar Islamic university. The driver behind coffee's diffusion in the Near East was its increasing adoption as a social beverage consumed outside religious ceremonies.

The Ottoman conquest of Egypt in 1516–17 facilitated

---

[1] Ralph Hattox, *Coffee and Coffeehouses: The Origins of a Social Beverage in the Medieval Near East* (Seattle, WA, 1985), p. 14.
[2] Ibid., p. 18.

its spread into the Turkishrun empire, reaching Damascus in 1534 and Istanbul in 1554. Two coffee houses were opened in the capital by Syrians, Hakam and Shem, from Damascus and Aleppo respectively. Situated in the city's downtown quarters, near the port and central markets, these coffee houses attracted an elite clientele, including poets who would try out their latest work on fellow literati, merchants engaged in games such as backgammon and chess, and Ottoman officials conversing with each other while seated upon luxurious couches and carpets. Such was Shems's success, he is said to have returned to Aleppo three years later, having made a profit of 5,000 gold pieces.

There was, however, a notable difference in colour between Arabian coffee and Turkish coffee. Arabic coffee (*qahwa*) was (and is) served as a semi-translucent, light brown liquid. The beans are lightly toasted before being cooled, crushed and mixed with spices such as ginger root, cinnamon and, especially, cardamom. The mixture is placed in a copper-bottomed pot, boiled with water for around fifteen minutes and then decanted into a smaller, pre-heated serving vessel (*dallah*) often with a long spout. The host pours a small cup, or *finjan*, for each guest.

The Ottoman Turks, by contrast, drank a dark, opaque beverage described by a contemporary poet as 'the negro enemy of sleep and love'—the forerunner of Turkish coffee, or *kahve*, today. The beans are blackened by roasting and then ground into a powder. The coffee is placed with water in a *cezve* (known outside Turkey as an ibrik or *briki*), a wide-bottomed open pot that narrows before reaching a broader rim. It is brought to the boil, removed from the flame and foam from the top of the liquid is spooned into the cups. The liquid may then be brought back to the boil (at least once, often twice), and additional liquor poured into the cups, while an attempt is made to retain the foam structure.

The practice of roasting beans was used by some Istanbul imams to argue that the consumption of coffee was illicit because the carbonization of the beans meant the drink was prepared from an inanimate (therefore forbidden) substance. In 1591 Bostanzade Mehmed Effendi, the Sheik ul-Islam (the highest religious authority), issued a *fatwa*, definitively declaring that the beverage remained of vegetable origin, as complete carbonization had not taken place. According to a contemporary chronicler:

> Among the *ulema*, the sheykhs, the viziers and the great, there was nobody left who did not drink it. It even reached such a point that the grand viziers built great coffee houses as investments, and began to rent them out at one or two gold pieces a day. [1]

The advent of the coffee house created possibilities for new forms of social interaction. Previously, entertaining others would have involved inviting them to one's house, providing a banquet, probably prepared by servants, and

---

[1]   Bernard Lewis, *Istanbul and the Civilization of the Ottoman Empire* (Norman, OK, 1963), p. 133.

entailing the display of possessions (and probably wife), all of which created a distinction between host and guest. Now one could meet peers at a coffee house, and exchange hospitality on a more equal footing through the simple expedient of buying each other cups of coffee. The layout of these early coffee houses facilitated an egalitarian ambience, as patrons were seated according to the order in which they arrived, at long benches or on divans running alongside the walls, rather than by their rank.

Alongside the elite, this format enabled those of lesser means to entertain each other and display their generosity. A visitor to Cairo in 1599 noticed:

> When soldiers go . . . into a coffee house and they have to get change for a gold coin, they will definitely spend it all. They regard it as improper to put the change in their pocket and leave. In other words, this is their manner of showing their grandiosity to the common people. But their grand patronage consists of treating each other to a cup of coffee, of impressing their friends with one cup of something four cups of which costs one *para* [penny]. [1]

So popular were coffee houses in Istanbul that it was claimed that in 1564, ten years after the first establishments opened, there were over fifty in operation; by 1595 this

---

[1] Hattox, *Coffee and Coffeehouses*, p. 99.

number had supposedly reached six hundred. [1] It seems likely that this figure involved some conflation of coffee houses with taverns and *boza* outlets, which may also have reflected the reality of establishments blurring the boundaries between activities. Coffee houses allowed the consumption of dubious substances and opportunities for gaming and gambling. By 1565 Suleiman the Magnificent, the sultan who had welcomed the first coffee houses to Istanbul, was issuing edicts to close the taverns, *boza* sellers and coffee houses of Aleppo and Damascus, where people 'continue to pass their time by amusing themselves and committing illicit and banned acts' that prevented them 'from carrying out their religious obligations'. [2] Further and more severe edicts were issued by his successors, Selim II (1566–74) and Murad III (1574–95).

These appear to have had limited impact, not least because the enforcing authorities and militia members were themselves usually patrons, and not infrequently proprietors, of these institutions. The coffee houses' success reflected a shift in the social and political structures of the Ottoman Empire. The centralized, hierarchical administration model gave way to a society in which power was fragmented, elites divided and religious and secular ideologies contested. The coffee house, where one could address anyone directly and engage in open conversation, became a symbol of this new

[1] Ayse Saracgil, 'Generi voluttari e ragion di stato', *Turcia*, 28 (1996), pp. 166–8.

[2] Ibid., p. 167.

culture.

Coffee houses were attacked by religious and political conservatives precisely because of their progressive connotations. Sultan Murad IV, who came to the throne in 1623 when still a minor, had great difficulty establishing his authority, and instituted a highly reactionary regime complete with networks of informers stalking the coffee shops and listening for gossip against him. In 1633, following a fire that destroyed five districts in Istanbul and was believed to have been started by the smoking of tobacco in a coffee house, Murad ordered the closure of all such establishments in the city. Orders were despatched to other cities in the empire, like the municipality of Eyüp, that

> required that, with the arrival of this order, persons are sent to destroy any coffee kilns that are in the zone that you govern, and that from now on no one should be permitted to open one. From now on anyone who opens a coffee shop should be strung up over its front door. [1]

Although the use of tobacco, introduced into Turkey at the turn of the seventeenth century, seems to have been the primary target, or pretext, for Murad's wrath (he is reputed to have stalked the city in disguise at night, exacting summary justice upon offenders), the problem for coffee shop owners, as one *pasha* pointed out, was that

---

[1] Ibid.

> In the coffee shops, the proprietors don't have the means to impose that customers, many of whom are soldiers, don't smoke; each one has his own tobacco in his pockets, he takes it out and smokes. Because (the smokers) have the privileges of state office holders, the proprietors of the coffee shops and the other inhabitants of the city cannot oppose them. [1]

The prohibition of coffee houses within Istanbul was still in place in the mid-1650s, though beyond the city walls coffee houses were trading openly, as they had probably continued to do in the empire's further regions throughout this period. By the last quarter of the seventeenth century, coffee houses had also reappeared in Istanbul, and travellers to the Ottoman territories remarked on the centrality of coffee houses in locations including Cairo's street markets, the caravan routes through the Arabian Peninsula and Istanbul's public gardens.

The spread of coffee throughout the Islamic world created a complex of long-distance trading networks that converged on Cairo, from where it was forwarded throughout the Ottoman Empire and eventually into Europe. Initially, wild coffee from Ethiopia was dried and shipped from Zeila (now in northern Somalia on the border with Djibouti). Here it would be added to the cargoes of spices

---

[1] Ibid.

originating in India and the Far East, carried up the Red Sea and offloaded at ports serving those regions where coffee had been adopted. The first coffee cargo mentioned comes in 1497 as part of a merchant's spice shipment from Tur at the Sinai peninsula's southern tip. [1]

Ethiopia remained the sole source of coffee until the 1540s, but a combination of rising demand and unreliable supplies due to conflicts between the African empire's Christian north and Muslim south led to coffee being cultivated in the highlands of the Yemeni interior between the coastal plain of Tihama and the capital city of Sana'a. Seeds from the small bean varieties found wild in Ethiopia were planted by peasants alongside subsistence crops on their family plots. These were the world's first coffee farms. The region remained the only centre of commercial coffee production for virtually two centuries. Villages of whitewashed stucco houses appeared throughout the mountains, surrounded by stone wall terraced plantations which were enriched with soil retrieved from the *wadis* following the rainy season. By the 1700s, these upland areas supported a population of some 1.5 million people.

The chain linking these producers to the final consumer was, as ever, a long and fragmented one. Transportation was exceptionally difficult, with nothing more than mule

---

[1] Michel Tuchscherer, 'Coffee in the Red Sea Area from the Sixteenth to the Nineteenth Century', in *The Global Coffee Economy in Africa, Asia, and Latin America, 1500–1989*, ed. William Gervase Clarence Smith and Steven Topik (Cambridge, 2003), p. 51.

paths connecting mountainous areas to lowland markets. Growers would bring their dried cherries to the nearest town to exchange them for goods like cloth and salt. The coffee would then pass through various intermediaries before ending up in the major wholesale market of Bayt al-Faqih located on the coastal plain. Here it was bought by merchants and held in warehouses, prior to being transported by camel train to the ports of Al-Makha (otherwise known as Al-Mocha, Al-Mokka and, to Europeans, Mocha) and Hudaydah for shipping. Most of these merchants were Banyans, members of a diaspora that spread from the Gujarat port of Surat to dominate trade around the Indian Ocean. They also controlled the Yemeni credit networks, making it likely they were the chief financiers, and effective initiators, of coffee cultivation.

Despite its fragmented nature, the coffee trade generated considerable revenue, particularly among the leaders of the Zaydi sects who commanded the loyalty of the interior tribes. This facilitated considerable resistance to Ottoman rule. They were forced out from Yemen in 1638 by the Qasimi dynasty of Zaydi imams, who united the country for the first time and gained control over Zeila, thus giving them an effective monopoly over the world's coffee supply from both Yemen and Ethiopia. Beans from both origins subsequently became known within the trade as 'Mocha', as they were exported together from the same port.

The success of the Zaydi uprising led to a reorganization of the Red Sea trade. Coffee destined for consumption within the empire was transported by dhow from Hudaydah to Jeddah. This was established as an obligatory entrepôt by the Ottomans, who used the generated revenue to support the holy places. Ships bringing down cereals from Suez returned loaded with coffee for Cairo. Here the city's merchants, who had begun regularly trading coffee as early as the 1560s, would despatch it to the empire's major Mediterranean centres such as Salonika, Istanbul and Tunis. After the 1650s, coffee was sent to Alexandria, where it was acquired by the Marseille traders who controlled access to western European ports.

Mocha meanwhile acted as the principal port to the rest of the coffee-consuming world—primarily the Islamic lands surrounding the Persian Gulf, Arabian Sea and Indian Ocean. As a result, it also became the leading entrepôt for Indian commerce throughout the Red Sea. The British East India Company opened a depot there as early as 1618 to get a stake in the trade, forwarding consignments of what was variously described as 'cowa', 'cowhe', 'cowha', 'cohoo', 'couha' and 'coffa' on to company factors (brokers) in Persia and Moghul India, over thirty years before coffee became available in Britain. Although European companies managed to divert a significant portion of the spice trade into their own hands during the seventeenth century as exchanges between Europe and Indochina increased, coffee remained largely concentrated within the Muslim mercantile networks.

Part of the problem for the Europeans was the continuing unpredictability of supply. The agriculture structure in the Yemeni highlands made it difficult for growers to respond to market demands. Jean de la Roque—author, traveller and son of the merchant who introduced coffee to Marseille—wrote accounts of two trading expeditions to Mocha from the Breton port of St Malo, in 1709 and 1711. These reveal it took six months to fill the ship's hold, even though the Frenchmen were using a Banyan broker, whose attempts to acquire beans on their behalf drove up the prices in Bayt al-Faqih. A Dutch factor they encountered reckoned on taking a year to acquire cargo for one voyage. By the 1720s, Red Sea coffee shipments had reached 12,000–15,000 tonnes per annum—effectively the world supply. [1] That volume remained largely unchanged over the next hundred years, even though by 1840 it accounted for no more than 3 per cent of world production. Given this, it is hardly surprising that, as they increasingly adopted the beverage, Europeans sought to establish alternative cultivation centres.

After the 1720s, the Dutch turned to Java and the French to the Caribbean, so their purchases from Mocha and Alexandria, respectively, declined. These were compensated by increased purchases by the British and the Americans. The revenues from the coffee trade were still such that Muhammad Ali, the expansionist ruler of Egypt, sought to conquer Yemen to bring them under his control. This led the British to seize Aden in 1839, protecting their influence

---

[1] Ibid., p. 55.

in the region, and establishing it as a free port in 1850. The absence of customs duties and the presence of deep-water quays and warehousing facilities saw Aden overtake Mocha as the region's chief coffee port. Today the harbour area of Mocha houses a small fishing fleet and many ruins, and is approached through silted-up channels, supposedly the consequence of nineteenth-century American ships discharging their ballast prior to taking coffee on board.

The main causes of decline within the Red Sea coffee economy were changes in taste among the overwhelmingly Muslim consumers. It was the turn to tea in India and Iran in the early nineteenth century that had the most dramatic impact, as these traditional Eastern markets were lost to coffee. In Egypt, tea was likely the more popular beverage, made from plants cultivated within the country. Part of Ataturk's programme for Turkey's modernization during the twentieth century's first half was converting it into a teadrinking country, substituting a beverage made from locally grown produce for an expensive import. It has taken the arrival of the Western coffee chains to stimulate a revival in Turkish coffee house culture.

Conversely, the one country where the coffee economy expanded over the last two centuries was Ethiopia. During the latter nineteenth century, Emperor Menelik used coffee export earnings to purchase firearms that were famously employed to defeat the Italians at Adowa in 1896, preserving Ethiopia's position as the lone independent African state after the continent's partition. As well as the 'wild' coffee from the southwest Oromo kingdoms such as Sidamo, Kaffa and Jimmah (probably produced on peasant plots to meet imperial demands for tribute), new plantations were established near the eastern region of Harar using cultivars of *Coffea arabica* that had evolved in growing areas around the world. These larger beans became known as Mocha Longberry, to distinguish them from the original Yemeni (and Ethiopian) 'Mocha'.

Coptic Christians in the north began to grow and consume coffee. The young Haile Selassie relied on coffee revenues to impose his authority in the 1930s. He, however, was unable to prevent the Italian Fascist occupation whose legacy includes the espresso bars of Addis Ababa and Asmara. Ethiopia remains one of the few growing countries that also consumes a significant portion (around 50 per cent) of its own coffee.

Chapter 3
Colonial Good

Few Europeans, except those under Ottoman rule, had tasted coffee before the middle of the seventeenth century. Its introduction into Europe led to the creation of the coffee house and café, whose appeal extended to large swathes of European society. The eighteenth century witnessed a dramatic reconfiguration of coffee's production centres as European states, such as the Dutch Republic, France and Britain, began growing coffee in their Asian and Caribbean colonial holdings to satisfy consumers' increasing demand.

Diffusion of Coffee Culture in Europe

This was not a straightforward story of Europeans being bowled over by the bean. Chocolate, coffee and tea came to the continent in quick succession, and consumer preferences shifted back and forth. Guild regulation complexities prevented traders from setting up premises to sell and serve coffee. Consequently, there were significant discontinuities in the diffusion of coffee culture. Venice was probably the first European city in which coffee was brewed, but a coffee house did not open there until a century afterwards. London was home to Europe's first coffee houses, yet the British were among the last and the least active of the European coffee producers. Conversely the French, late converts from chocolate, went on to dominate both consumption and colonial production during the eighteenth century.

Diffusion of Coffee Culture in Europe [1]

	First Record of Coffee in Territory	First Commercial Shipment Received	First Coffee House Opened	First Colonial Plantings
Italian States	1575 Venice	1624 Venice	1683 Venice	
Netherlands	1596 Leiden 1616 Amsterdam	1640 Amsterdam	1665 Amsterdam 1670 The Hague	1696 Java 1712 Suriname
England	1637 Oxford	1657 London	1650? Oxford 1652 London	1730 Jamaica
France	1644 Marseilles	1660 Marseilles	1670 Marseilles 1671 Paris	1715 Reunion 1723 Martinique
German States		1669 Bremen	1673 Bremen 1721 Berlin	
Habsburg Empire	1665 Vienna		1685 Vienna	

---

[1] Data derived from literature survey.

Coffee's adoption across Christian Europe reflected the continent's complex relationship with the Islamic Near East. Outbreaks of fascination with the 'Orient' provoked interest in coffee, yet travellers writing in the early seventeenth century often sought to rescue the beverage from its Muslim associations by reimagining its past. The Italian Pietro della Valle suggested coffee was the basis of nepenthe, the stimulant prepared by Helen in Homer's *Odyssey*. The Englishman Sir Henry Blount claimed it was the Spartans' black broth drunk before battles. By locating coffee among the ancient Greeks, they effectively claimed it for European civilization, and reminded contemporaries of coffee-drinking Christians within the Ottoman borders. There is, though, no evidence that Pope Clemente VIII tasted coffee and baptized it as a Christian beverage in the 1600s, although the story's widespread circulation suggests those with a stake in the coffee trade wished he had done so.

Coffee was present in Venice in 1575, as the coffeemaking equipment recorded in the inventory of a murdered Turkish merchant in the city confirmed. By 1624 it was being shipped into the city for sale by apothecaries as a medicinal product, and in 1645 a shop selling beans appears to have been licensed. [1] Coffee's use spread to other Italian states: Tuscany awarded a monopoly for trading in coffee in 1665. Regulation protecting the apothecary trade probably account for the late appearance of the first café allowed to serve coffee in Venice

---

[1] Markman Ellis, *The Coffee House: A Cultural History* (London, 2004), p. 82.

in 1683. By 1759 the city authorities were forced to cap the number of cafés at 204—a limit breached within four years.

It seems likely that similar restrictions obscured coffee's early history in the Habsburg Empire, particularly in territories bordering on, and often invaded by, the Ottomans. A Turkish delegation despatched to Vienna in 1665 to ratify a peace treaty included two men charged with coffee preparation. By 1666, when the delegation left, there was supposedly a thriving domestic coffee trade. It was dominated by Armenians, such as Johannes Diodato, who in 1685 was awarded the first licence for preparing and selling coffee—that is, operating a coffee house. [1]

This contradicts the embroidered story that Georg Franz Kolschitzky, a spy behind Ottoman lines during the 1683 siege of Vienna, had obtained the sacks of coffee beans abandoned by the retreating Turks as his reward, and used them to introduce coffee into the city. Supposedly Kolschitzky, in Turkish dress, started by hawking pre-brewed coffee around the city, while petitioning authorities to be allowed to open his own shop. When that was granted, he set up the famous coffee house at the sign of the Blue Bottle. In 1697, three years after Kolschitzky's death, the licensed guild Bruderschaft der Kaffeesieder (Brotherhood of Coffeemakers) was established. [2] Their key innovation

---

[1]   Bennet Alan Weinberg and Bonnie K. Bealer, *The World of Caffeine* (New York, 2002), pp. 74–9.

[2]   Karl Teply, *Die Einführung des Kaffees in Wien* (Vienna, 1980); Andreas Weigl, 'Vom Kaffeehaus zum Beisl', in *Die Revolution am Esstisch*, ed. Hans Jürgen Teuteberg (Stuttgart, 2004), p. 180.

was adding milk to the coffee, with customers using a colour chart to indicate their desired shade. This was the origin of the *Kapuziner*, a beverage the colour of the Capuchin monks' tunic. As well as sweetening (or perhaps masking) the taste, milk symbolically transformed the black Muslim brew into a white Christian confection.

England evolved the first European coffee house culture. The leading figures in bringing the bean to Britain were also émigrés from the Ottoman Empire. Nathaniel Conopios, a Greek student at Balliol College, Oxford, was the first person recorded drinking coffee in England in May 1637. A Jewish manservant from the Levant named Jacob has sometimes been credited with opening a coffee house in the same city in 1650, but, if he existed, he probably served, rather than sold, coffee to his master's companions. There is no doubt, however, that Pasqua Rosee, an ethnic Armenian from the Ottoman city of Smyrna (now Izmir) opened London's, and Europe's, first documented coffee house sometime between 1652 and 1654. Such was the new institution's swift take-off, there were 82 coffee-house keepers registered in 1663 with the City of London authorities.

Rosee's business began as a stall in St Michael's churchyard in the heart of the City of London, the independent borough at the metropolis's centre that contained most of London's financial and commercial institutions. Merchants would come from the nearby Royal Exchange to continue conversations while sipping coffee under the awning of Rosee's stall. According to the first reference to the business, in 1654, it served 'a Turkish-kind of drink made of water and some berry or Turkish-beane (that was) somewhat hot and unpleasant (but had) a good after relish and caused some breaking of wind in abundance'. [1]

A handbill extolled *The Vertue of the Coffee drink. First publiquely made and sold in England by Pasqua Rosee*. Coffee, it observed, was 'a simple innocent thing' prepared by being

> ground to Powder and boiled up with Spring water, and about a half pint of it to be drunk . . . as hot as possibly can be endured . . . It will prevent Drowsiness and make one fit for business, if one have occasion to Watch and therefore you are not to Drink of it after Supper, unless you intend to be watchful, for it will hinder sleep for 3 or 4 hours. It is observed in Turkey, where this is generally drunk, that they are not troubled with the Stone, Gout, Dropsie or Scurvey, and their Skins are exceedingly clear and white. It is neither Laxative nor Restringent. [2]

The beverage's principal benefit was the physiological effect of 'promoting watchfulness'. The advantage over drinking the 'small beer', or weak ale—the traditional form of refreshment in a city where water supplies were often filthy—can easily be gauged. The coffee house quickly took

---

[1]   Ellis, *The Coffee House*, p. 33.
[2]   *The Vertue of the Coffee Drink* (London, undated, possibly 1656), now in the British Library.

over from the tavern as the principal public venue in which to conduct business.

Local tavern-keepers resented Rosee's success, which they complained was stealing their business; but as he was not trading in alcohol, they were unable to argue that he was infringing on their licences. Instead they challenged his right to trade on the grounds that he was not a citizen, resulting in Rosee entering into partnership with Christopher Bowman, a member of the Company of Grocers. Together they transferred the enterprise into a set of rooms overlooking the churchyard. It continued trading under a sign of Rosee's silhouette, known as the Turk's Head, although there is no record of his being involved in the business beyond 1658. Bowman died in 1662, after which his widow took over until the premises were destroyed in the Great Fire of 1666.

It was no accident that coffee houses were established during the Cromwellian era following the English Civil War's end. The guilds' power weakened and the prevalent cultural values of egalitarianism and sobriety suited the introduction of an alcohol-free venue for socializing in which customers were treated as equals. Early proprietors installed long tables, seating everyone and anyone without a suggestion of hierarchy. The coffee was brewed over a fire and decanted into coffee pots, from which waiters poured it into bowls—known as dishes—for the customers.

Coffee houses survived the monarchy's restoration in 1660 because Royalist opponents of the Parliamentary regime had also taken advantage of the opportunities these venues created for unmonitored conversation. When the Earl of Clarendon proposed to the Privy Council in 1666 closing down coffee houses, he was reminded by William Coventry that 'in Cromwell's time . . . the King's friends had used more liberty of Speech in these Places than They durst do in any other'. [1] That was true in Oxford where the first documented coffee house outside the capital was opened by apothecary Arthur Tillyard in 1656. Tillyard was 'encouraged to do so by some royalists, now living in Oxon, and by others who esteemed themselves either virtuosi or wits'. [2]

The term 'virtuosi' described gentlemen possessed of intellectual curiosity about cultural novelties, rarities and the fledgling field of empirical, quasi-scientific enquiry associated with figures such as Francis Bacon. As virtuosi were not courtiers, they were free to learn about new phenomena and discuss them within the so-called 'penny universities'–the price coffee houses charged for a dish of coffee.

Tillyard's customers included Issac Newton, the father of modern physics, the astronomer Edmond Halley (of comet fame) and the collector Hans Sloane, whose bequest formed the basis of the British Museum. Most of the virtuosi were not such outstanding scholars but enthusiasts who could be enticed into coffee houses to inspect displays of curiosities. Don Saltero's coffee house, opened in London in 1729 by James Salter, included attractions like 'Painted

---

[1] Ellis, *The Coffee House*, p. 73.
[2] Brian Cowan, *The Social Life of Coffee: The Emergence of the British Coffeehouse* (New Haven, CT, and London, 2005), p. 90.

Ribbands from Jerusalem with the Pillar to which our Saviour was tied when scourged' and 'A large Snake 17-foot-long, taken in a pigeon house in Sumatra, it had in its belly 15 fowls and 5 pigeons'. [1]

Other great patrons of coffee houses were City businessmen, who congregated in certain establishments to conduct their affairs. The most well known are Lloyd's, founded in 1688, which became the centre for maritime insurance, and Jonathan's, whose role as a rudimentary stock exchange saw it play a leading role in the South Sea Bubble, the frenzied share speculation and subsequent market crash, of 1711. The value of networking at the coffee house was recognized by Samuel Pepys, who took the decision to frequent them over taverns in 1663, drinking coffee until he 'was almost sick', but acquiring considerable wealth from the kick-backs on his naval supply deals. [2]

A distinction needs to be drawn between the success of the coffee houses and that of coffee. Chocolate and tea were available as alternative beverages within the coffee houses as early as 1660. In 1664 the Grecian coffee house, home to the newly founded Royal Society, advertised that it not only sold chocolate and tea but offered customers lessons in how to make them. Tellingly, a second aborted attempt by King Charles's ministers to suppress coffee houses in 1675 defined them as any house selling 'Coffee, Chocolate, Sherbett, or

---

[1] *A Catalogue of the Rarities to Be Seen in Don Saltero's Coffee House in Chelsea* (London, 1731).

[2] Samuel Pepys, diary entry, Friday 23 January 1663, www.pepysdiary.com.

Tea'. The fact that Whites, the first 'chocolate house', was only founded in 1693, and Thomas Twining opened his original teahouse in 1711, is not so much an indication of the later take-up of these beverages as of their earlier availability.

Coffee's association with the coffee house may have held back its adoption in the home. The coffee house was essentially a male environment in which talking to strangers was encouraged. The only women present were either serving or 'servicing' the customer's needs. The *Women's Petition against Coffee*—a 1674 condemnation of both coffee and coffee houses on the grounds that they kept men away from the home and rendered them impotent—was probably sponsored by brewers keen to recapture lost customers, but it played on this gender division.

Well-bred women were directed towards tea. Tea was favoured by several royal role models, notably the Portuguese Catherine of Braganza, who introduced it to the English court in 1662 when she married Charles II. Subsequently England was ruled by two sovereign queens, Mary (1689–94) and her sister Anne (1702–14), both of whom were tea drinkers. Women might take tea together, either at home, or publicly in tea gardens where the open-air settings conferred a visibility, rendering them respectable places for ladies.

It is difficult to gauge the full extent of London's coffee house explosion. The next apparently accurate figures are from Henry Maitland, whose exhaustive survey of the capital in 1739 revealed 551 coffee houses—around one per thousand head of population. In the City borough alone,

144 coffee houses were recorded—a number roughly equal to taverns and inns. The 8,000-plus gin palaces Maitland also identified in London, outnumbering coffee houses by some eighty to one in the poorest quarters of the capital, were, however, an indication of coffee's elite beverage status. Tea consumption began to permeate into the lower classes after the 1740s. Tariffs were cut on the Chinese tea imported by the British East India Company, but coffee bought on the international market continued to incur heavy duties.

In the second half of the eighteenth century coffee houses began to offer alcohol alongside coffee, effectively turning back into taverns, as the number of pubs with names like The Turk's Head indicates. One example was The Turk's Head in Gerrard Street, London, which hosted a literary club in 1764. Members included the great lexicographer Dr Samuel Johnson and his biographer James Boswell. They drank tea and wine respectively. Some coffee houses doubled as bordellos, as seen in Hogarth's 1738 picture *Morning*, depicting King's Coffee House in London's Covent Garden. Those hosting business activities at times turned into exchanges themselves (Lloyd's), while meeting places for the virtuosi increasingly became private gentlemen's clubs to preserve societal distinction. A commercial guide to London published in 1815 listed just twelve coffee houses.

This contrasted with the French café, which, though slower to become established, developed into a social institution appealing to all classes during the eighteenth century. Coffee was traded in Marseilles by the 1640s but remained largely unknown in Paris until 1669. In that year a diplomatic mission was sent by Sultan Mehmed IV to Louis XIV, probably instigated by the French ambassador to Constantinople to impress his own sovereign. The delegation remained for almost a year, entertaining influential courtiers with Turkish delicacies like coffee in a mansion refurbished as a Persian palace. This inspired 'Turkomania' among French society's upper echelons, as satirized in Molière's *Le Bourgeois gentilhomme*. Vendors started selling coffee during trade fairs in the Saint-Germain commercial district, and an Armenian named Pascal established the first Parisian coffee house in 1671, only to see it fail after this Turkish fad subsided.

It was not until the early 1700s that a long-running dispute between the guilds of the *limonadiers* (distillers), grocers and apothecaries over the right to sell coffee was finally resolved. The *limonadiers* were granted an effective monopoly over the service of decoctions—be they lemonade, coffee or gin—to seated customers on their premises. This created the café format as an establishment in which coffee and alcohol were taken—ooften with the intention of using the properties of the former to counteract those of the latter.

An early example was the Café Procope, established in 1686 by the Italian-born *limonadier* Francesco Procopio, who renamed himself François Procope. It was furnished with gilded mirrors, marble tables, painted ceilings and chandeliers, recalling an aristocratic salon rather than a sultan's seraglio. Coffee and its accompaniments were served using porcelain tableware and silver cutlery. Located across from the newly founded Comédie Française, its elite patrons could enjoy encountering theatrical performers, while feeling safe in their surroundings. The Procope became the model for the grand cafés throughout eighteenth-century Europe, such as Florian's in Venice and Caffè Greco in Rome.

By 1720 there were in the region of 280 cafés in Paris, rising to around 1,000 in 1750 and 1,800 in 1790, serving a population of approximately 650,000. Most of these catered to more modest clienteles than the Procope, providing Parisians with venues for meeting and socializing, playing games such as chequers, or gambling on the various lotteries. Clay pipes were kept ready loaded for the use of customers, as smoking tobacco was almost ubiquitous. The fittings and furnishings were matched to the clienteles' social station, while cafés seeking extra space spilled onto the boulevards. Locations immediately outside city limits had lower rents, offering customers access to pastoral pleasures during the day, and the demi-monde by night.

The café was part of the masculine world. Although many cafés were run by couples, with the woman working front of house while her male partner prepared the drinks and accompaniments in the backroom 'laboratory', few women set foot in a café for fear of being mistaken as prostitutes due to the café's public nature and its trade in alcohol. If women were served coffee, it was likely taken to their carriage to be drunk in privacy.

Bourgeois women chose chocolate, not least for its

supposedly medicinal qualities. Coffee began to challenge this primacy with the spread of *café au lait*. As Philippe Dufour explained in his 1684 book on coffee, tea and chocolate, 'when [ground] coffee is boiled in milk and a little thickened, it approaches the flavour of chocolate which nearly everyone finds good.' [1] Furthermore, *café au lait* could be presented as French rather than foreign in origin. As an enthusiastic noblewoman states in 1690, 'We have here good milk and good cows; we've taken it into our heads to skim the cream . . . and mix it with sugar and good coffee.' [2]

The Low Countries witnessed an even more rapid adoption of coffee among both sexes and throughout the classes. Coffee-making equipment was frequently found among probate inventories of lower-class and middling households in eighteenth-century Amsterdam. As early as 1726 it was claimed coffee 'has broken through so generally in our land that maids and seamstresses now have to have their coffee in the morning or they cannot put their thread through the eye of their needle'. [3] It seems weavers, undertaking piecework within their homes, were fuelling themselves with sugarsweetened coffee to avoid leaving their looms.

---

[1] Philippe Sylvestre Dufour, *Traitez nouveaux et curieux du café, du thé et du chocolat*, 3rd edn (The Hague, 1693), p. 135.
[2] Julia Landweber, 'Domesticating the Queen of Beans', *World History Bulletin*, XXVI/1 (2010), p. 11.
[3] Anne McCants, 'Poor Consumers as Global Consumers: The Diffusion of Tea and Coffee Drinking in the Eighteenth Century', *Economic History Review*, LXI, S1 (2008), p. 177.

Coffee's growing popularity led European trading companies to try and secure supplies: the situation became acute in 1707 when the Ottoman administration imposed an export ban on coffee outside the empire. By then Nicolaes Witsen, a governor of the Dutch East India Company (VOC), had already started to plant coffee on Java in 1696. The seeds came from Malabar in India, where legend has it that coffee was planted by the Muslim scholar Baba Budan, who smuggled the seeds back in his clothing following a pilgrimage to Mecca. A more likely explanation for the presence of coffee in Malabar was that this was an outgrowth of the Banyan coffee trade.

On Java, the VOC operated by coercing indigenous chiefs into supplying a fixed quantity of coffee in exchange for a low, pre-established, price. Local lords required peasant households under their control to provide them with coffee as part of their feudal obligations. Coffee was a crop that conferred neither financial nor nutritional value to themselves, so the peasants had little incentive to improve cultivation techniques. They preferred to meet quotas by growing coffee in their domestic plots or forest gardens. Peasant households in western Java, where cultivation was concentrated, were at times forced to relocate to locations suitable for growing coffee in plantations, under the control of the lord's subordinates. [1]

[1]   M. R. Fernando, 'Coffee Cultivation in Java', in *The Global Coffee Economy in Africa, Asia and Latin America, 1500–1989*, ed. William Gervase Clarence Smith and Steven Topik (Cambridge, 2003), pp. 157–72.

Regular shipments from Java to Holland began in 1711, enabling Amsterdam to establish the first European coffee exchange. In 1721, 90 per cent of the coffee on the Amsterdam market originated in the Yemen; by 1726, 90 per cent was supplied from Java. [1] Deliveries from the island continued to increase until the middle of the century, but tailed off as new plantations in the Caribbean took over.

The Dutch were partly responsible for this. In 1712 they introduced coffee to Suriname, a colonial enclave on the northeastern coastline of mainland Latin America, bordering the Caribbean Sea. Exports began in 1721 and surpassed those from Java by the 1740s. In Suriname, cultivators had no option but to produce coffee—the crop was grown on plantations tended by slave labour.

In 1715 the French planted coffee on the island of Bourbon (now Réunion), part of the Mascareignes archipelago off Africa's east coast. In the 1640s the French East India Company had begun colonizing the uninhabited island, granting concessions of land to French settlers who worked them with African slaves. Arabica trees obtained from Yemen, in defiance of the Ottoman ban, proved so successful that the Company decreed in 1724 that it would repossess any concessions that failed to plant coffee, and even debated imposing the death penalty for deliberately

---

[1] Steven Topik, 'The Integration of the World Coffee Market', in *The Global Coffee Economy*, p. 28.

damaging a coffee tree. [1]

In the 1720s the French also introduced coffee to their Caribbean territories, starting with Martinique. Gabriel de Clieu, a young naval officer, supposedly transported cuttings of coffee plants from Paris's botanic gardens to the island in 1723, many years later publishing a heroic account of how he shared his water ration with them during the voyage. It now seems that planting began in 1724 using seeds from Bourbon and Suriname.

When and how coffee arrived on Saint Domingue (now Haiti) is less clear, but its success soon outstripped production everywhere else in the Caribbean. The French obtained the colony at the end of the Nine Years War in 1697 when Hispaniola was divided into two. Santo Domingo, the eastern portion (today's Dominican Republic) remained under Spanish control, while Saint Domingue occupied the rugged western third of the island. As elsewhere in the Caribbean, the lower coastal areas were devoted to sugar-cane plantations, while coffee farms were established in the mountainous interior.

Up until the 1730s, the French East India Company refused to allow coffee grown in either Bourbon or the Caribbean to be sold in France. This was to protect its own monopoly on the more highly priced Mocha. Instead, coffee from these origins was shipped to the Amsterdam exchange—including *café marron*, a coffee species now known

---

[1] Gwyn Campbell, 'The Origins and Development of Coffee Production in Réunion and Madagascar', in *The Global Coffee Economy*, p. 68.

as Mauritian coffee (*Coffea mauritiana*) found growing wild on Bourbon. It proved inferior to the cultivated Arabica and was abandoned in the 1720s. By the 1750s, the proportion of coffee from the Americas traded on the Amsterdam exchange matched that from Asia.

The influx of colonial coffee into France after the import ban was lifted in the mid-seventeenth century drove down the price and made the beverage more accessible to the lower classes. A degree of snobbery developed around coffeedrinking styles, with the philosopher Jacques-François Demachy drawing a comparison in 1775 between

> a woman of high society, comfortably settled in her armchair, who consumes a succulent breakfast to which mocca has added its perfume for a well-varnished tea table, in a . . . gilded porcelain cup, with well-refined sugar and good cream; and . . . a vegetable seller soaking a bad penny loaf in a detestable liquor, which she has been told is Café au Lait, in a ghastly earthenware pot. [1]

By the 1780s, 80 per cent of the world's coffee supply came from the Caribbean, principally Saint Domingue. More plantations were established between the 1760s and 1780s, until the value of coffee exports matched those of sugar cane. The colony's success lay in the low costs of production—achieved principally with imported African slave labour.

---

[1] Emma Spary, *Eating the Enlightenment* (Chicago, IL, 2012), p. 91.

A remarkable, if disturbing, guide to coffee growing in Saint Domingue was published by the planter P. J. Laborie in 1798. He described all the stages of coffee cultivation, from clearing the land to bagging the beans. The book includes an account of the innovative 'West Indian process' for pulping the cherries, using a water-channels system to soften the fruit and pass it through a series of graters. Yet what leaps out is his belief that 'the negroe' (his term for slaves) 'is that creature that we are forced to keep in his natural state of thraldom to obtain from him the requisite services; because . . . under a different condition he would not labour'.[1]

When buying slaves, Laborie advised looking for features such as an open cheerful countenance, a clean and lively eye, sound teeth, sinewy arms, dry and large hands, strong loins and haunches, and an easy and free movement of the limbs. After purchase, they were forced to drink 'sudorific potions' for a fortnight to sweat out diseases picked up on the voyage, and the 'unpleasant but necessary' act of branding them was performed.

New slaves had to be 'seasoned'—introduced gradually into farm work as they assimilated to the cooler climate. Laborie preferred to buy young boys and girls of around fifteen who could be formed to 'the Master's own ideas' while undertaking gardening and weeding. They would then join the main gang, working on the plantation from sunrise to sunset, under the authority of a driver—an entrusted slave

[1]   P. J. Laborie, *The Coffee Planter of Saint Domingo* (London, 1798), p. 158.

equipped with a whip.

Maintaining authority was a priority. Insubordination such as talking back to master or driver was more severely punished than any offence committed by one slave against another—including violent assaults and rape. Laborie wrote of the need to clean whips between floggings to avoid spreading infection.

Racial politics in Saint Domingue was complicated. Over a third of coffee plantations, and a quarter of all the slaves, were owned by so-called *gens de couleur*. This group comprised French settlers' mixed-race offspring recognized by their fathers, plus, a growing number of black former slaves, who had been freed by their masters. By 1789 the colony had 28,000 *gens de couleur* and 30,000 whites, but both groups were 'outnumbered' by the 465,000 slaves.

The 1789 French revolution encouraged the *gens de couleur* to assert their rights to be treated as equal to the white population, while slaves used the instability to stage their own rebellions for better conditions. From 1791 these forces came together in an uneasy alliance led by Toussaint L'Ouverture, a freed black slave who had at one point owned a coffee plantation and fifteen slaves. A sickening cycle of violence, foreign intervention, repression and war lasted until 1804 when Saint Domingue declared its independence, renamed itself Haiti and abolished slavery. Over a thousand coffee plantations were destroyed, including that of Laborie, who fled to Jamaica. Although new farms were established, the coffee trade was effectively lost because European states and the USA shunned Haiti for fear of legitimizing black rule.

Within Europe, the coffee supply disruption was intensified by the British navy's blockade of French territories. Napoleon's response was to encourage home-grown chicory as a substitute. Chicory was also promoted by Frederick the Great in Prussia. He employed so-called 'coffee sniffers' in the 1780s to clamp down on consumption. The practice of bulking out the beans with roasted chicory became pervasive. Even in the early twentieth century, William Ukers, founding editor of the trade journal *Tea and Coffee*, complained that many Europeans had 'acquired a chicory and coffee taste such that is doubtful if they would appreciate a real cup of coffee should they ever meet it'. [1]

Nonetheless the first half of the nineteenth century saw a continued increase of coffee consumption throughout Europe. Swedish novels of the period feature scenes in which coffee is drunk by all classes: for example, in Emilie Flygare-Carlén's *Pål Värning*, published in 1844, the hero is a poor fisherman who makes a hazardous journey to buy coffee for his sick mother. At the tavern-cum-shop he encounters an elderly maid, for whom 'sitting by the kitchen stove with a pipe in her mouth, and with the coffee pot on the fire was . . . the finest pleasure in life'. Coffee came to be sold in 'Colonial Goods' shops—aptly named, as most European supplies still came from their imperial possessions.

The first drip-brewing apparatus appeared in the early

---

[1] W. H. Ukers, *All About Coffee* (New York, 1935), p. 554.

nineteenth century, so-called de Belloy pots named after the coffee-loving Archbishop of Paris. A filter compartment containing the ground coffee separated two chambers, so that hot water poured into the upper could filter down into the lower. Later pots were designed so the water could be heated on the stove, then the apparatus flipped over to allow percolation to take place. [1] Over the course of the century, fashionable forms of equipment such as syphon systems and hydrostatic percolators found favour among the elite, while drip brewing became widespread across much of Europe.

The demise of Saint Domingue sparked a revival of coffee production in Asia. Java's popularity saw the island's name adopted as a synonym for coffee in the United States. However, what was sold as 'Java' was as likely to have originated in Sumatra and other Dutch colonies across the Indonesian archipelago. It could take five months for coffee to be shipped to New York, during which time the beans aged and often turned brown due to sweating. The coffee became prized for its low acidity, and continued to be shipped in sailing vessels even after the advent of steam.

The Dutch colonial authorities continued to work through local rulers, introducing the so-called Collection System that required peasant households to set aside a portion of their land or labour to cultivate commercial crops sold exclusively to the state. The autobiographical novel *Max Havelaar*, penned by a former administrator in 1860,

---

[1]   Enrico Maltoni and Mauro Carli, *Coffeemakers* (Rimini, 2013).

showed how peasants starved while the Dutch indulged their indigent lords.[1] By the 1880s, 60 per cent of Java's peasant households were forced to grow coffee. Tending to the trees took up 15 per cent of their time, yet generated only 4 per cent of their income, due to the low fixed prices.

The British also expanded their colonial coffee production, most notably on Ceylon (Sri Lanka). Having gained control of the coast from the Dutch during the Napoleonic Wars, they set about conquering the interior, overthrowing the independent Kingdom of Kandy in 1815. British entrepreneurs cleared the forests to set up coffee plantations, killing off many of the island's elephants and importing workers from the heavily indebted Tamil population of the Indian region of Madras. Untold numbers died 'on the road' to these plantations or due to working conditions when they got there.[2]

By the late 1860s total British coffee production in Ceylon and India was approaching that of the Dutch colonies. In 1869 an outbreak of leaf rust caused by the fungus *Hemileia vastatrix* began. By the mid-1880s the coffee plantations had been largely destroyed, and were converted to growing tea, cementing the triumph of leaf over bean in Britain. By 1913 Ceylon was a net coffee importer.

The rust outbreak spread throughout Asia, wiping

[1] Multatuli, *Max Havelaar: Or the Coffee Auctions of a Dutch Trading Company* [1860] (London, 1987).
[2] Donovan Moldrich, *Bitter Berry Bondage: The Nineteenth Century Coffee Workers of Sri Lanka* (Pelawatta, Sri Lanka, 2016).

out most of the production in Java, Sumatra and the rest of the East Indies, as well as India. It even travelled as far as Africa and the Pacific Islands. Some planters substituted Arabica trees with Liberia's native species, *Coffea liberica*. Its harshtasting beans found little favour, except among locals in Malaysia and the Philippines, where it became the basis for Barako, a dark-roasted, highly caffeinated coffee. In any case Liberica, too, proved susceptible to the fungus. By the outbreak of the First World War, Asia provided just one-twentieth of the world coffee supply, compared to around one-third before the rust appeared. The global coffee economy was now centred in the Americas.

Chapter 4
Industrial Product

Coffee was transformed into an industrial product during the latter part of the nineteenth century by two nations in the Americas: Brazil and the United States. Brazil's ability to rapidly expand coffee output without significantly raising its prices enabled the U.S. to absorb this into its enlarging consumer economy. Brazil extended the coffee frontier into its hinterlands by replacing a slave labour force with imported European peasant labourers. U.S. consumption per capita tripled between the mid-nineteenth and mid-twentieth centuries, as consumers moved from home roasting to purchasing pre-prepared, branded industrial coffee products. Once Central America and Colombia began to compete for the U.S. market, new forms of coffee politics appeared, as states strove to protect their national interests.

Coffee in the USA:
From the Colonial Era to the Civil War

Americans' preference for coffee is often presented as

an outcome of the struggle for independence. Tea became a symbolic focus for the colonists' demands for 'no taxation without representation'. The British government imposed a duty on tea imports into the colonies, which were also part of an East India Company monopoly. Protestors staged the Boston Tea Party on 16 December 1773, tipping tea chests off ships in Chesapeake Bay harbour. Consequently, the story goes, American patriots switched to coffee.

United States Coffee Consumption Statistics, 1880–1950 [1]

Date	Total Imports (million lbs)	Consumption per Capita (lbs)	Share of World Imports (%)
1800	8.8	1.65	
1830	38.3	2.98	
1860	182.0	5.78	28.7
1890	490.1	8.31	36.1
1920	1,244.9	11.88	56.1
1950	2,427.7	16.04	63.6

The reality is more complicated. Coffee had long been available in the colonies; particularly in Boston, where Dorothy

[1] Derived from William H. Ukers, *All About Coffee* (New York, 1935), p. 529; Mario Samper, 'Appendix: Historical Statistics of Coffee Production and Trade from 1700 to 1960', in *The Global Coffee Economy in Africa, Asia, and Latin America, 1500–1989*, ed. William Gervase Clarence Smith and Steven Topik (Cambridge, 2003), pp. 419, 442–4.

Jones became the first person licensed to sell 'coffee and cuchaletto [chocolate]' in 1670. Coffee houses spread through the city, mostly doubling as taverns: the Green Dragon, founded in 1697, was a regular meeting place for political activists. Coffee imports into the colonies were also controlled by the British, however, coming principally from Jamaica.

After the Tea Party, the patriotic response was to secure alternatives to British supplies. John Adams requested 'a Dish of Tea providing it has been honestly smuggled or paid no Duties' in 1774. In 1777, after his wife Abigail described how Bostonian women had broken into a warehouse in search of coffee and sugar, Adams hoped that 'females will leave off their attachment to coffee', and start drinking beverages made from American- grown products. [1]

Coffee gained popularity once French supplies from Saint Domingue began to arrive into the newly independent United States. By 1800 consumption was over 680 grams (1.5 lb) per capita. [2] The U.S. developed a lucrative re-export trade in coffee during the Napoleonic wars. Coffee from the Caribbean was carried to Europe in American vessels to avoid naval blockades.

After 1820 consumption increased significantly, sparked by a fall in prices from 21 cents a pound in 1821 to 8 cents in 1830. The cause of the fall was speculators hoarding coffee

[1] Steven Topik and Michelle McDonald, 'Why Americans Drink Coffee', in *Coffee: A Comprehensive Guide to the Bean, the Beverage and the Industry*, ed. R. Thurston, J. Morris and S. Steiman (Lanham, MD, 2013), p. 236.
[2] Figures derived from William H. Ukers, *All About Coffee*, p. 529.

in anticipation of a Franco-Spanish war that didn't break out, leaving them to dump beans on the international market. As the world supply expanded, coffee prices rarely exceeded 10 cents per pound for the following two decades. The U.S. federal government removed import taxes on coffee in 1832, and by 1850 consumption was over 2.3 kilograms (5 lb) per capita.

Cuba became the United States' primary supplier following the demise of Saint Domingue, with many American investors acquiring plantations on the island. After a series of natural disasters destroyed hundreds of trees in the 1840s, however, many switched to sugar. Thereafter the U.S. increasingly obtained low-price coffee from Latin America, especially Brazil.

The Civil War (1860–65) was a pivotal event in the United States' coffee history. Union troops were plied with coffee: a daily ration of about 43 grams (1.5 oz) of coffee a day totalled a staggering 16 kilograms (36 lb) a year. That could easily support the consumption of ten cups of coffee a day. Generals, aware of caffeine's psychoactive effects, ensured their men had drunk plenty of coffee before battle; some soldiers carried grinders fitted into the butt of their rifles. The Union blockade of the Southern coastline meant the Confederate states—and their troops—were forced to use substitutes like chicory and acorns.

Coffee's centrality to the troops' existence can be gauged from the fact that the word 'coffee' appears more frequently in soldiers' diaries of the period than 'rifle', 'cannon' or 'bullet'. Sergeants distributing coffee rations avoided accusations of favouritism by facing the other way when calling up men to receive their allotment. John Billings, an artilleryman, in his memoir *Hardtack and Coffee,* described how dipping the former in the latter killed off the weevils in the biscuit as they floated to the surface. He recalled:

> If a march was ordered at midnight . . . it must be preceded by a pot of coffee; if a halt was ordered in mid-forenoon or afternoon, the same dish was inevitable. . . . It was coffee *at* meals and *between* meals . . . and today the old soldiers who can stand it are the hardest coffee drinkers in the community. [1]

The bloodiest day of the Civil War was 17 September 1862 at Antietam. Nineteen-year-old Sergeant William McKinley (later a U.S. president) passed along the front line serving the troops coffee, despite coming under heavy fire. The effect on morale 'was like putting a new regiment in the fight' according to their commanding officer. [2]

Foundations of the Coffee Industry

When Civil War soldiers returned home, their coffee-

---

[1] John D. Billings, *Hardtack and Coffee* (Boston, MA, 1887), pp. 129–30.
[2] Jon Grinspan, 'How Coffee Fueled the Civil War', www.nytimes.com, 9 July 2014.

drinking habit stimulated the emerging domestic coffee industry. By the 1880s America was importing one-third of the world's coffee, occasioning the establishment of the New York Coffee Exchange in 1882.

Throughout the nineteenth century, most coffee in the largely rural U.S. was purchased in bulk as green beans from a catalogue supplier or general store. Batches of beans were roasted at home in a pan over a wood stove, stirred for around twenty minutes. Better-off households might have a sealed home roaster turned by hand or by steam. Home grinders were becoming common by mid-century, but the mortar and pestle were frequently used to crush roasted beans into powder.

Preparation techniques were simple—coffee grounds were heated with water in a kettle. Household guides recommended boiling for 20–25 minutes. A variety of additives were employed to encourage the grounds to settle to the bottom—most frequently egg white, but also isinglass (a fish-based gelatine product).

The first popular refinement was the Old Dominion coffee pot in 1859. This was an early percolator, in which coffee was placed in a perforated container within a lower chamber which boiled the water, while a condenser unit above liquefied and recycled escaping vapour. Users were advised to leave coffee and water in the pot on the stove overnight, then boil again for ten to fifteen minutes before breakfast. The thin-bodied yet bitter-tasting brew became the characteristic taste of American coffee.

By the 1840s the emergence of major urban centres created conditions for new wholesale coffee-roasting businesses. These supplied stores with loose, ready-roasted beans sold by weight. Beans were roasted using the pull-out roaster patented by James W. Carter of Boston in 1846. This comprised a long roasting cylinder set into a brick-built furnace fired by coal. A pulley system drew the cylinder in and out of the furnace, and it was filled and emptied using sliding doors set in its sides.

Operators judged when coffee was ready from the colour of the smoke emitting around the cylinder's edges. They unloaded the hot coffee into trays then stirred the beans by hand until cool. Some simply dumped the hot coffee directly onto the floor, spreading it with rakes and sprinkling with water. One observer recalled how 'the contact of water and hot coffee caused so much steam that the roasting room was in a dense fog for several minutes after each batch of coffee was withdrawn from the fire'. [1]

In 1864 Jabez Burns patented his self-emptying roaster. Within the revolving cylinder set in a brick oven, a so-called 'double screw' allowed the beans to be moved uniformly up and down for an even roast. The key was that the beans could be emptied from the front, into a cooling tray, without taking the cylinder from the fire. Burns developed further refinements for cooling and grinding the coffee that saved time, significantly reducing the price difference between

---

[1] Ukers, *All About Coffee*, p. 589.

green and roasted coffee at wholesale and retail. In 1874 Burns declared:

> It is preposterous to suppose that household roasting will be continued long in any part of this country, if coffee properly prepared can be had . . . It will never pay for small stores to roast if the large manufactories do the work well . . . By doing the work with proper care they will not only secure . . . large sales for themselves, but will command the roasting for other parties. [1]

Coffee in the United States was set to become a massmanufactured, industrial product—branded and marketed to an emergent consumer society.

The Rise of the Coffee Brands

John Arbuckle of Pittsburgh, who ran a wholesale grocery business with his brother, was one of the first purchasers of the new Burns machine. In 1865 he started selling roasted coffee in reinforced paper packaging (developed for peanuts). Three years later he patented an egg and sugar glaze for roasting, claiming it prevented the beans from staling by protecting the surface from air and clarifying the brew. His advertisements showed a woman roasting

---

[1] Ibid., p. 596.

beans at a wood stove and lamenting, 'Oh, I have burnt my coffee again', and being advised by her well-dressed guest to 'Buy Arbuckles' Roasted, as I do, and you will have no trouble.' The text underneath stated, 'You cannot roast coffee properly yourself.'[1]

In 1873 Arbuckles launched Ariosa, which became the first nationally known coffee brand: glazed beans wrapped in a distinctive yellow packaging with Arbuckles in red, and a flying angel trademark design above the name. By 1881 the company was roasting with 85 Burns machines in New York and Pittsburgh factories, and had distribution depots in Chicago and Kansas.

Ariosa's most dedicated market was the cowboys, ranchers and pioneers in the Far West. Many were demobilized Civil War soldiers who had acquired a taste for coffee. Each pack contained a peppermint stick, the sweet taste of which was designed to offset that of the coffee. Wagon train cooks allegedly called 'Who wants the candy?' to entice volunteers to grind the beans. Each pack contained coupons that could be redeemed for items such as tools, guns, razors, curtains and even wedding rings. The angel image was used to persuade Native Americans that coffee could confer spiritual powers, experienced as a caffeine buzz.

The rise of wholesale coffee roasting saw the development of several other prominent brands. Jim Folger set up Folgers coffee roasting company in San Francisco during the 1850s

---

[1] Mark Prendergast, *Uncommon Grounds* (New York, 2010), p. 49.

Gold Rush. In 1878 Caleb Chase and James Sanborn merged their coffee companies, begun in Boston, and started the Seal brand—the first to use sealed cans for packing.

Canned coffee became the U.S. standard, although the process also sealed in air, so staling remained an issue. Hills Brothers, another San Francisco company, addressed this by introducing vacuum-packed coffee in 1900. This technology favoured pre-ground coffee, like Hills' top-of-the-range Red Can brand. In 1892 the Cheek-Neal company introduced Maxwell House—named after a swanky Nashville hotel that they supplied.

By 1915, 85 per cent of consumers preferred to purchase pre-packaged branded coffee over loose roasted beans. Some 3,500 brands existed, though not all were on local grocery stores' shelves. Around 60 per cent of the market was with door-to-door delivery companies: the largest, the Jewell Tea Company, earned half its income from coffee sales. Chain stores' own-brand coffee accounted for another significant market share. The Great Atlantic and Pacific Tea Company, commonly known as the A&P, sold their own brand Eight O'Clock Coffee—adding 'theatre' by installing in-store grinders.

Coffee cemented its position as America's national beverage during the early twentieth century as consumption reached 5 kilograms (11 lb) per capita. The United States now imported well over half of the world's coffee supply, and roasters positioned their brands as inherently American with names such as 'Buffalo' and 'Dining Car Special'. Thomas

Wood & Co. boasted that its Uncle Sam's Coffee came from 'his own possessions in Porto Rico, Hawaii and Manilla [*sic*]'.

Most roasters were reticent about the origins of their blends, however. Hills Brothers trademarked the figure of an Arab in a flowing robe in 1897, using brand names like Caravan, Santola, Timingo and Saxon, which obscured more than they revealed. Java and Mocha remained the only acknowledged production sources, promoting cowboy slang for coffee—jamoka'. Arbuckles warned consumers to 'beware of buying low-grade package coffee falsely purporting to be made of Mocha, Java and Rio; this being a cheap device employed by the manufacturers to deceive unwary customers.' [1] Ariosa was widely presumed to be composed of beans from Rio and Santos. By the mid-1870s more than 75 per cent of the coffee consumed in the United States came from Brazil.

Coffee in Brazil

Coffee was supposedly introduced to the Portuguese colony by Francisco de Melo Palheta in 1727. The story goes that the diplomat Palheta was sent to resolve a dispute between Dutch and French colonies in Guiana. He returned to Brazil with seeds hidden in a bouquet from his lover, the French governor's wife. He planted them in Para, his home

---

[1] Ibid., p. 71.

region, but until 1822 coffee remained a minor crop in Brazil compared to sugar.

Coffee's fortunes were transformed when it was introduced into the mountainous Paraíba valley region, south of Rio de Janeiro, in response to rising prices following the demise of Saint Domingue. Coffee trees took well to the *terra roxa*—the well-drained and nutrient-rich red clay soil found in Brazil's central-southern states.

Cultivation techniques were crude, with little regard for the environment. Hillside forests were cut and burnt down, creating a layer of fertilized ash above the soil into which the seedlings were set. Planting did not consider soil erosion, and bushes grew in full sun, sucking out the ground's goodness. Production was augmented by bringing more virgin land into the system.

Brazilian Coffee Production Statistics, 1870–1990 [1]

Date (two-year averages)	Brazilian Production (million sacks)	World Production (million sacks)	Share of World Production (%)
1870–71	3.1	6.6	46.9
1900–1901	14.5	18.7	77.5
1930–31	25.1	37.0	67.8
1960–61	32.9	68.9	47.7
1990–91	28.5	98.4	28.9

[1] Data from Francisco Vidal Luna and Herbert S. Klein, *The Economic and Social History of Brazil since 1889* (Cambridge, 2014), pp. 355–9.

The large estates (*fazendas*) owned by the wealthy elite used slave labour. Each slave might tend 4,000–7,000 plants. Little maintenance was carried out. The natural drying process was used, prior to hulling and despatching to Rio on mule trains. Lack of soil maintenance contributed to the Rio beans' poor reputation, being prone to mould and off-flavours. Today, 'Rio-y' still describes such defects.

After the U.S. banned imports of slaves in 1807, North American slave traders shifted to the Brazilian market, setting up a triangular exchange: American goods traded into Africa, in exchange for slaves who were sold in Brazil, to buy coffee for delivery back to the U.S. This lasted until 1850 when the British ended the Atlantic slave trade through direct naval intervention.

Existing slaves (around one-third of the population) remained central to the Brazilian economy. An internal slave market developed with southern Brazilian coffee planters buying slaves from the north. Only in 1871 was the so-called 'Law of the Free Womb' passed, making children of slaves free from birth, followed in 1888 by the 'Golden Law' freeing all remaining slaves.

In 1872 Brazil's imperial family was overthrown. A new republic was created, dominated by the *Paulistas*, the coffee barons of São Paulo state.

Dominance of São Paulo, City and State

The *Paulistas* replaced slave labour forces with poor European immigrants, known as *colonos*. They worked for wages on large coffee estates, but were given housing and a small plot to grow their own food. In 1884 the Brazilian state began subsidizing initial costs of transporting migrants, and by 1903 over 2 million had arrived. Over half came from Italy, attracted by the promise of land, but found they effectively became indentured labourers, required to pay back the cost of their voyage. The terms were so harsh that the Italian government banned subsidized migration schemes in 1902. Portugal and Spain then became principal sources of Brazilian *colonos*.

Coffee production rose phenomenally in this era—from 5.5 million sacks in 1890 to 16.3 million in 1901. Brazil accounted for 73 per cent of world coffee output between 1901 and 1905. Most was grown in the São Paulo region, where over 500 million coffee trees had been planted by 1900, meaning this one state alone produced nearly half of the coffee grown throughout the world.

The dramatic increase was achieved by bringing more and more land under cultivation. The coffee frontier spread south and west across São Paulo, moving through the central highlands into the state's hinterlands. Trading shifted to the port of Santos, assisted by São Paulo's extensive railroad development, including lines solely for transporting coffee.

The 1905 agricultural census captured the characteristics

of the São Paulo coffee economy. Sixty-five per cent of the workforce on the 21,000 coffee farms in the state was foreignborn. The top 20 per cent of farmers controlled 83 per cent of the land, produced 75 per cent of the coffee and employed 67 per cent of the agricultural labour force. The biggest producer, the German-born Francesco Schmidt, owned 7 million coffee trees and employed over 4,000 workers.

The *Paulistas'* agriculture system did not create a coffee monoculture, however. It was common for *colonos* to grow food crops among coffee trees, and many *fazendas* practised mixed farming. The state of São Paulo was self-sufficient in food.[1]

Valorization

Brazil's world coffee market dominance reached its zenith in 1906, when it produced 20.2 million bags of coffee, around 85 per cent of the total world output. This bumper crop forced a change in the country's coffee strategy. During the nineteenth century Brazil had increased its coffee revenues by expanding production while maintaining low wholesale prices that stimulated demand. At the turn of the century, however, supply overtook demand, and prices

---

[1] Francisco Vidal Luna, Herbert S. Klein and William Summerhill, 'The Characteristics of Coffee Production and Agriculture in the State of Sao Paolo in 1905', *Agricultural History*, XC/1 (2016), pp. 22–50.

plummeted from 13 to 6 U.S. cents per pound.

In 1906 the São Paulo state government subsidized a consortium of bankers and brokers led by Hermann Sielcken, a German-American coffee merchant, to buy up the coffee surplus and keep it off the market. By 1910 prices recovered to over 10 cents per pound and most of the syndicate's holdings were sold off by the end of 1913.

This 'valorization' of the coffee price orchestrated by Brazilian authorities was a significant moment in coffee's history: it was the first time producer countries had dictated trade terms to consumer nations. It caused outrage in the U.S., where Sielcken was hauled before a congressional committee in 1912. His explanation that there would have been a revolution in São Paulo without this scheme met with an unsympathetic response: 'Do you think that would have been a worse condition than that we [U.S.] should pay 14 cents a pound?'[1]

Valorization was regularly used by Brazilian authorities to regulate the amount of coffee on the world market thereafter, maintaining export prices at over 20 cents per pound. The São Paulo state started an agency to organize coffee interests, which was subsequently transformed into the national Instituto Brasileiro do Café (IBC).

The 1930s Great Depression destroyed this progress. The problem was a massive supply increase caused by new land being brought into production, with bumper crops being

---

[1] Prendergast, *Uncommon Grounds*, p. 84.

recorded every other year from 1927 onwards. The Brazilian harvest alone exceeded world demand in these years. By 1930 Brazil held 26 million bags in reserve stocks. Prices crashed to under 10 cents for the rest of the decade.

The IBC managed what became a desperate response by the authorities. Between 1931 and 1939 a network of 75 huge incineration plants was established and 80 million bags of coffee (three years' global supply) went up in smoke. Tax penalties on new plantings were introduced and alternative uses for coffee were found, including making coffee bricks to fuel trains. The ibc tried to promote consumption through advertising and by opening Brazilian coffee houses in Europe, Russia and Japan.

Central America

Another cause of Brazil's problems was that its dominance of world supply had diminished with the emergence of Colombia and the Central America states—Costa Rica, El Salvador, Guatemala, Honduras, Nicaragua and Panama, plus Mexico. Until 1914 Brazil provided 75 per cent of U.S. coffee imports; between the First and Second World Wars, this fell to around 50 per cent. Furthermore, coffees from these other states enjoyed a significant price premium on the U.S. exchanges.

That premium derived from their coffees' superior cup quality, a reflection of the greater care in cultivation and harvesting, and using wet processing. This was undertaken at washing stations, or *beneficios*, which became the central point in connecting cultivators to the market. Large estates often operated their own processing plants, but small producers usually sold their cherries directly to the *beneficio*. *Beneficios* often extended credit to planters, effectively tying them into supplier deals. This placed *beneficio* operators in a strong position to ensure they received good-quality, ripe cherries, requiring selective picking throughout the season.

Central American states needed export earnings and encouraged their coffee frontiers into remote uncultivated, though often populated, highland regions. Conversion to commercial coffee growing required parcelling up land into private holdings, and creating a workforce with a sufficient stake to produce coffee to the quality levels required. Much of the year-round cultivation was undertaken by peasants working small-scale, often household-production units, whether as independent owners or within a variety of tenancy agreements. [1] This still left the issue of where to recruit pickers, however.

A range of solutions emerged according to circumstance. In El Salvador, vagrancy laws were used to force native populations from their lands and turn them into labourers on plantation-style estates. This generated a class of coffee oligarchs, effectively controlling the country, creating

---

[1] William Roseberry, 'Introduction', in *Coffee, Society and Power in Latin America*, ed. W. Roseberry, L. Gudmondson and M. Samper Kutschbach (Baltimore, MD, 1995), p. 30.

inequalities and conflicts that persisted throughout the twentieth century. In 1932 a revolt of impoverished coffee workers resulted in the *Matanza* (massacre) of tens of thousands of indigenous Salvadorans by government forces.

Conversely, in Costa Rica, the government passed homesteading laws, allowing settlers to claim titles to unoccupied land on the high plateaus where few native inhabitants were to be found. These settlers established independent smallholdings with backing from the *beneficios*, who in turn operated on credit supplied by importing firms mainly based in London, which functioned as an entrepôt for Costa Rican coffee.

Guatemala became the first Central American state to make a significant impression in the global market, becoming the world's fourth-largest coffee exporter by the end of the nineteenth century. Under the Liberal Premier General Barrios, in the 1870s, foreign purchasers could acquire large estates, tempted in by adverts in European papers such as *Le Monde*. Coffee growers exploited laws allowing departmental governors to compel villages to supply labourers, securing themselves seasonal workers for harvesting. German nationals were attracted into the country, and by the early twentieth century owned 10 per cent of the coffee farms (*fincas*), processed 40 per cent of the coffee harvest and controlled 80 per cent of the country's exports.

The outbreak of the First World War severely disrupted the European coffee market. This intensified a reorientation of Central American exports towards the United States, which began when San Francisco broker Clarence Bickford started sample cuppings with his buyers in the 1900s. These demonstrated that classifying beans by their colour and size alone (as done on the New York exchange) was inadequate for determining quality. Small-sized beans such as Guatemala's had been traded at a discount—now they enjoyed a premium. [1]

San Francisco's port and improved rail links made it a distribution hub for Central American coffee throughout the U.S. The opening of the Panama Canal in 1914 facilitated connections between coffee-exporting Pacific seaports of Central and Latin America and receiving ports in the North American and European markets. In 1913 U.S. imports of coffee from Central America totalled 36.3 million pounds; in 1918 they reached 195.3 million.

Colombia

After the First World War, Colombia emerged as the world's second-largest coffee producer—increasing output from 61,000 metric tons in 1913 to 101,000 in 1919 and 256,000 in 1938.

Coffee was supposedly introduced to Colombia by Jesuit priests, some of whom required their parishioners to plant coffee trees as an act of penance. Production

---

[1] Ukers, *All About Coffee*, p. 424.

was established on the mountainsides of the Andes' three branches (*cordilleras*), from north to south. Difficult terrain made railroad construction uneconomical, so coffee was brought by mule train to the Magdalena and Cauca rivers for shipment to the Caribbean ports of Barranquilla and Cartagena, or conveyed using an aerial cable car-style system to Buenaventura on the Pacific coast.

Commercial cultivation expanded in the late nineteenth century. Bogotá and Medellín merchants invested in coffee *haciendas* in the three departments of Santander, Cundinamarca and Antioquia. Inspired by Guatemala's success, they introduced similar techniques of planting shade plants to avoid soil erosion, selective picking of ripe cherries and wet processing. They relied principally on family units for production under different tenure systems—sharecropping in Santander, tenant farming in Antioquia and extensive estates known as *latifundia* in Cundinamarca.

During the 1920s Colombia's coffee output doubled, responding to the high prices obtained via the Brazilian valorization schemes and Colombia's quality premium of over 20 per cent. The coffee frontier moved south into Caldas, Tolima and Huila. Coffee accounted for 60–80 per cent of the country's exports, but left the Colombian industry highly exposed to the Brazilian bumper harvests and subsequent price collapse during the Great Depression. Social conflict ensued as landowners attempted to pass on their losses by altering contract terms with cultivators. Many disputes descended into violence, notably on the *latifundia*.

At this juncture, the Colombian state stepped in, founding the Federación Nacional de Caféteros de Colombia (National Federation of Colombian Coffee Growers) in 1927 to act as 'a private entity carrying out essential public functions for the national interest'. [1] Funded by the introduction of a levy on every bag of coffee leaving the country, its remit was to manage the country's coffee policy in the 'best interests' of its growers. As well as providing educational, financial and technical services to its members, the FNCC regulates the country's exports sector and promotes Colombian coffee abroad.

Its wide-ranging powers enabled the FNCC to effectively manipulate the price premium between Colombian and Brazilian coffee. During the 1930s it deliberately lowered the differential to obtain greater market share in the United States. By 1937 Colombia had captured 25 per cent of the U.S. market.

Inter-American Coffee Agreement

The collapse of the coffee price during the Depression forced the leading Latin American producers to start negotiating with each other to find solutions to the crisis. In 1936 they established the Pan-American Coffee Bureau to promote consumption in the United States, while the

[1]   Marco Palacios, *Coffee in Colombia, 1850–1970* (Cambridge, 1980), p. 217.

Brazilian ibc and Colombian FNCC entered into a price maintenance agreement, which quickly fell apart, with Brazil accusing Colombia of 'free-riding' on Brazilian efforts to regulate the coffee supply by withholding stocks from the market. In 1938 Brazil flooded the market with coffee in frustration, driving prices back down, but following the outbreak of war in Europe in 1939, it became imperative for producers to find a way to avert any further price collapse.

On 28 November 1940, the Inter-American Coffee Agreement was signed by all fourteen coffee-producing states in the western hemisphere along with the United States, which recognized the value of ensuring the stability of supply. The agreement stated, 'It is necessary and desirable to take steps to promote the orderly marketing of coffee, with a view to assuring terms of trade equitable for both producers and consumers by adjusting supply to demand.'[1]

The national agencies representing the coffee producers negotiated quotas for their exports to the United States, with the system entering into force in April 1941. By the end of the year prices had doubled, and they remained strong thereafter.

Creating Consumers

Consumption levels in the United States rose steadily

---

[1] Paul C. Daniels, 'The Inter-American Coffee Agreement', *Law and Contemporary Problems*, 8 (1941), p. 720.

throughout the first half of the twentieth century. Average annual coffee imports into the U.S. doubled between 1915–20 and 1946–50. Even the Great Depression failed to halt this progress. By 1939 coffee had become an everyday household good—98 per cent of U.S. households reported drinking it.

Coffee's fortunes were favoured by broader developments in American society. Prohibition between 1920 and 1933 saw cafés supplant saloons, so coffee rather than alcohol became the legally sanctioned beverage for socializing outside the home. An increasing emphasis on a light lunch during the working day resulted in more daytime consumption of coffee. Even so, the principal venue for coffee consumption was the home.

But What Did the Consumer Want?

A 1924 survey for J. Walter Thompson established that 87 per cent of housewives cited flavour as the most important factor in their choice of blend. However, 'it is extremely difficult for the average person to make clear distinctions where flavour is concerned'. [1]

The Texan retailer Harry Longe said the market was segmented into four types of purchasers—all housewives—and came up with 'Any Blend' messages to appeal to each of them. [2]

---

[1] Prendergast, *Uncommon Grounds*, p. 157.
[2] Ukers, *All About Coffee*, p. 484.

The 'Know-it-all-about-Coffee' who cannot
find anything to suit her cultivated taste:
IMPROVE THE COFFEE AND
YOU IMPROVE THE MEAL

The corner of the table that holds the coffee pot is the balancing point of your dinner. If the coffee is a 'little off' for some reason or other—probably it's the coffee's own fault—things don't seem as good as they might; but when it is 'up to taste' the meal is a pleasure from start to finish. If the 'balancing point' is giving you trouble, let *ANY BLEND* coffee properly regulate it for you.

The bride of a few months who knows very little about coffee, but
wants to find a good blend that
she and her husband can rely on:

A SUCCESSFUL SELECTION

of the coffee that goes into the every-morning cup will arrive on the day when *ANY BLEND* is first purchased. Many homes have been without a success for a long time, but of course, they didn't know of *ANY BLEND* —and even now it is hard to really know *ANY BLEND* until you try it. That is why we seem to insist that you ask for an introduction by ordering a pound.

Those satisfied with their existing coffee:
A SERVICE THAT SAVES

is the serving of *ANY BLEND*, when coffee is desired.

ANY BLEND saves many things. It saves worry, for it is always uniform in flavour and strength. It saves time, for when you order *ANY BLEND* we grind it just as fine or as coarse as your percolator or pot demands. *ANY BLEND* also saves expense, because there is no waste, as you know just how much to use, every time, to make a certain number of cups.

And for households with staff:

CAN YOU NAME YOUR COFFEE?

or is it one of those many unknown brands that comes from the store at the order of your cook? Let the cook do the ordering, for you are lucky if you have one you can rely upon, but tell her you prefer *ANY BLEND* to the No-Name blend you may now be using. *ANY BLEND* has one distinct advantage over all others; it is freshly roasted. Tell the kitchen-lady, now, to order *ANY BLEND*.

Longe's words played on consumers' lack of confidence about coffee. Coffee was regarded as representing the household to outsiders, so he created anxiety about its quality. Getting it right was presented as vital to domestic harmony— no new bride wants to live in a 'home without success'.

Brands and Advertising

By the end of the 1930s, over 90 per cent of roasted coffee was purchased in pre-weighed, trademarked packages. There were over 5,000 coffee brands, but the three leading players—A&P, Maxwell House, and Chase & Sanborn—held 40 per cent of the market. Their dominance was partially due to over half of all purchases being made in grocery chain outlets, such as those operated by A&P. By 1929 Maxwell House and Chase & Sanborn had been acquired by General Foods and Standard Brands, respectively—two huge corporations who utilized their economic power to ensure the brands were given prominence on supermarket shelves.

Manufacturers began persuasive communications campaigns, utilizing the mass channels that developed in the interwar era. Advertising agencies produced campaigns like Maxwell House's advertisements in glossy magazines, featuring the original sophisticated hotel and President Teddy Roosevelt's alleged endorsement that the coffee was 'good to the last drop'. A sponsored radio variety programme—*The Maxwell House Show Boat*—was introduced in 1933, and soon became the country's top show. Hollywood celebrities sipped coffee while chatting to presenters between music and acts, while listeners were reminded that 'your ticket of admission is just your loyalty to Maxwell House coffee.' Within a year of the show's launch, sales had risen by 85 per cent. [1]

---

[1] Prendergast, *Uncommon Grounds*, pp. 193–6.

Much of the major roasters' messaging played on the sense of unease around coffee identified by Longe. Chase & Sanborn regularly ran advertisements where wives were reproached by their husbands for failing to serve satisfactory coffee. Such adverts were meant to be educational, urging readers to purchase 'dated' coffee (stamped with the day of store delivery) and vacuum packs, to ensure freshness. Even then, Hills Brothers, institutors of the process, disclaimed, 'The coffee is turned over to you in perfect condition. Here our responsibility ceases, and unless you will cooperate with us by seeing that the coffee is made properly, our efforts and your money will be wasted.'[1]

War and After

The U.S. entry into the Second World War saw a brief period of rationing between 1942 and 1943, but wartime experiences further increased coffee's popularity.

Soldiers proved avid consumers, encouraged by officers who realized the value for morale. As in the Civil War, coffee seems to have served as stimulant and comfort, and perhaps a relief from monotony. An early post-war study of the Navy suggested that, while at sea, sailors consumed twice as much as civilians—even onshore personnel drank 50 per cent more

---

[1] Steve Lanford and Robert Mills, *Hills Bros. Coffee Can Chronology Field Guide* (Fairbanks, AK, 2006), pp. 19–25.

than the national average. [1]

Munitions workers proved more productive when allowed the new 'coffee breaks'. These were introduced throughout the military. The practice spread into post-war civilian life, with around 60 per cent of factories adopting it by the mid-1950s. This was partly a consequence of the Pan-American Coffee Bureau's heavy promotional campaign in favour of workplace coffee breaks. It also advocated 'coffee breaks on the road', arguing that coffee kept drivers alert in an increasingly motorized America.

A survey in winter 1954 found that consumers drank an average of two and a half cups a day. Two cups were drunk at home—usually at breakfast and dinner—with the remainder taken either in cafés/restaurants or at work. City dwellers enjoyed 2.8 cups per day, and those in rural districts averaged 2.3 cups. The highest levels of consumption, however, were found in the Midwestern farming belt, perhaps reflecting the Scandinavian origins of many of its inhabitants.

The End of an Era

Immediately following the Second World War, U.S. consumption levels per capita reached an all-time peak of over 8.6 kilograms (19 lb) per person for those over ten years old. Latin America was producing 85 per cent of the world's

[1] Andrés Uribe, *Brown Gold: The Amazing Story of Coffee* (New York, 1954), pp. 42–4.

output and sending 70 per cent of it to the U.S., where coffee was now consumed in virtually every household. The concept of the American 'cup of Joe'—a term for 'ordinary coffee' that first appeared in the 1930s—was firmly established. This presented as a thin-bodied, weak-flavoured coffee served in a comparatively large volume to accompany meals. Its taste profile reflected the blandness of the Brazilian beans at its base, the over-extracted coffee that resulted from brewing with a percolator, and the parsimoniousness of American housewives with the quantities of coffee they used.

By the end of the 1950s, however, it was already clear that consumption levels in the U.S. were declining as the younger generation turned to soft drinks, with Europe on the verge of overtaking North America as the leading consumer continent. Latin American producers, meanwhile, were again suffering low prices caused by oversupply, exacerbated by the rise of new players in Africa and Asia growing cheaper Robusta. Coffee had become a global commodity.

Chapter 5
Global Commodity

Coffee became a global commodity during the second half of the twentieth century. The foundation was the planting of Robusta as a hardier alternative to Arabica, reviving coffee production in Africa and Asia. Its cheaper price facilitated everyday coffee drinking among new consumers, and dramatically altered the beverage's taste and forms. International institutions developed to regulate the world coffee market, but proved incapable of protecting producers from price volatility, culminating in the coffee crisis at the century's end.

Robusta and the African Revival

Robusta varieties from the Belgian Congo were introduced into the Dutch East Indies during the 1900s to replace Arabica plants lost to coffee rust. By the 1930s over 90 per cent of the East Indies output was Robusta. This found a ready market among American roasters, enabling them to advertise that their blends contained Java or Sumatra.

The Second World War and subsequent Independence wars destroyed much Indonesian coffee production. It was not until the 1980s that Indonesia again became the world's largest Robusta producer.

Leading Producer States by Decade [1]

1960s	1970s	1980s	1990s	2000s	2010s
Brazil	Brazil	Brazil	Brazil	Brazil	Brazil
Colombia	Colombia	Colombia	Colombia	Vietnam	Vietnam
Angola	Ivory Coast	Indonesia	Indonesia	Colombia	Colombia
Uganda	Mexico	Mexico	Vietnam	Ethiopia	Indonesia
Ivory Coast	Indonesia	Ivory Coast	Guatemala	India	Ethiopia
Coast	Ethiopia	Ethiopia	India	Mexico	India

Robusta returned Africa to a central role within the global coffee economy that it had not known since the decline of Mocha. By 1965 the continent accounted for 23 per cent of world production compared to 2 per cent in 1914. [2] Seventy-five per cent of this output was Robusta, principally grown in the former French and Belgian colonies of West and Central Africa, as well as Uganda and Angola.

The Ivory Coast raised production from under 16,000

---

[1] Data courtesy ICO, 1965/6 to 2016–17. Note 1960s = crop years 1965–6 to 1969–70; 2010s = crop years 2010/11 to 2016/17.

[2] Stuart McCook, 'The Ecology of Taste', in *Coffee: A Comprehensive Guide*, ed. R. Thurston, J. Morris and S. Steiman (Lanham, MD, 2013), p. 253.

tonnes in 1939 to 114,000 tonnes in 1958. Production expanded rapidly after independence in 1960, reaching 279,500 tonnes in 1970. Levels maintained over the next twenty years. During the 1970s, the Ivory Coast was the third-largest coffee producer in the world (after Brazil and Colombia), and the leading Robusta exporter. Credit is due to Félix Houphouët-Boigny, the country's first president.

Houphouët-Boigny was a coffee farmer who campaigned against the French plantation owners' privileges in the colonial era, particularly their exploitation of involuntary labour. Once these advantages were removed, native growers proved more efficient. After independence, Houphouët-Boigny urged his compatriots not to 'vegetate in bamboo huts' but to concentrate on growing good coffee to 'become rich'. [1] He retained the French *Caisse de stabilization*—the 'stabilization fund' agency. This set prices for buying and selling coffee at all stages. It connected producers to processors and exporters. Although trade remained in private hands, the government protected growers by offering guaranteed prices, thereby encouraging new land into production in the country's central forest zones.

The *Caisse* system was used by nearly all the Francophone African states. It was financed through a tax applied on exporters' profits. The aim was to accumulate revenues during high world prices to maintain the returns to producers when prices were low, and to invest in raising

---

[1]   Jennifer A. Widner, 'The Origins of Agricultural Policy in Cote d'Ivoire', *Journal of Development Studies*, XXIX/4 (1993), pp. 25–59.

productivity and agricultural diversification. This worked comparatively well in the Ivory Coast. Farmers received an average of 70 per cent of the world price between 1974 and 1982. The temptation to drive down the internal prices paid to producers to generate surpluses to support non-economic government projects proved too much for many states, however, and the *Caisse* frequently became a font of institutional corruption.

States formerly under British rule maintained the marketing boards introduced by colonial authorities. These purchased processed beans for export, and returned the exchange received to the state. The marketing boards undertook sorting, grading and blending. They rewarded quality production through the differential prices they paid for coffee that could be included in premium lots.

In Kenya, white settlers, most famously Karen Blixen, set up farms in the early twentieth century. They planted Arabica, but used Bourbon varieties rather than the Typica that originated in Ethiopia. This partly explains the contrast between the floral, citrus-like flavours found in Ethiopian coffees, and the jammy, blackberry taste of many Kenyan origins. In the 1950s, following the Mau Mau uprising, agricultural reforms were introduced that encouraged the establishment of family holdings combining subsistence farming with the planting of cash crops, notably coffee. After independence in 1963, the Kenyan government retained the Coffee Board's central auction system, whereby exporters purchased lots classified according to cup characteristics, and growers received the average price for their class, thereby rewarding quality. By contrast, in newly independent Tanzania, coffee was sold by the Coffee Board in homogeneous lots, and quickly lost its reputation, which only began to be restored following reforms in the mid-1990s.

Uganda meanwhile became a leading volume producer of Robusta, increasing production from 31,000 tonnes in the late 1940s to 119,000 by independence in 1962. In 1969 this peaked at 247,000 tonnes, most of which was grown by smallholders on garden plots in areas like the Lake Victoria Crescent. Unusually, the Ugandan producers washed their Robusta, raising its quality. Production began to fall from the mid-1970s because of the Amin dictatorship's disastrous policies and the subsequent decade of political and military instability. The Coffee Marketing Board grew into a bloated bureaucracy, returning less than 20 per cent of their coffee's market price to the farmers, despite coffee being responsible for over 90 per cent of the country's export earnings. [1]

Instant Coffee

In 1929 the Brazilian authorities, desperate for alternative uses for their surplus beans, approached the Swiss multinational

---

[1] Moses Masiga and Alice Ruhweza, 'Commodity Revenue Management: Coffee and Cotton in Uganda', *International Institute for Sustainable Development* (2007).

food manufacturer Nestlé to ask if it could develop a coffee stock cube. It took Nestlé research scientist Max Morgenthaler over six years to come up with a palatable soluble coffee, by which time both Brazilian interest, and his own research team, had long been withdrawn. [1]

Nescafé, a spray-dried extract, was launched in 1938. War in Europe led to concentration of production in the U.S., where the War Department bought up virtually all production for military use. It returned to Europe in GIS' backpacks and the CARE (Cooperative for American Remittances to Europe) packages despatched at the war's end. In 1965 freeze-dried Nescafé Gold Blend was launched as a premium product, packaged in jars.

The major American roasters also started producing soluble coffees. Maxwell House's version overtook Nescafé in the U.S. in 1953. By the end of the decade solubles had a 20 per cent market share, primarily at the low-price end of the grocery sector. They, like Nescafé, used blends composed of upwards of 50 per cent Robusta beans.

Instant coffee came to define national taste preferences and consumption practices within underdeveloped markets. In the tea-dominated uk, coffee consumption doubled during the 1950s, coinciding with the launch of commercial television. Viewers found advertising breaks gave them sufficient time to make a cup of instant coffee, but not to brew a traditional cup of tea. By the 1990s coffee was

---

[1] Nestlé, *Over a Cup of Coffee* (Vevey, 2013), pp. 25–30.

outselling tea in terms of value (not volume), with instant coffee comprising 90 per cent of sales. Nescafé's iconic 'Gold Blend Couple' TV campaign, launched in 1987—featuring a will-they-won't-they relationship between two neighbours—entrenched its position as the market leader. Over half the uk population tuned in to watch the final episode in December 1992, and the campaign is claimed to have increased sales by 70 per cent. [1]

At a trade fair in Thessaloniki, Greece, in 1957, a Nescafé representative mixed instant powder and cold water in a cocoa drink shaker, creating a thick foam. Diluted with more water and served over ice, it proved very refreshing. The company began promoting this new use, which was adopted by young Greeks, becoming a symbol of the outdoor lifestyle. Frappé became Greece's national summer beverage. [2]

Emergence of European Coffee Styles

The advent of Robusta significantly changed the profiles of other national tastes in post-war Europe. By 1960, 75 per cent of coffee consumed in France was Robusta, necessitating the use of dark caramelized roasts to counteract the beans'

---

[1] Claire Beal, 'Should the Gold Blend Couple Get Back Together?', www.independent.co.uk, 28 April 2010.

[2] Vivian Constantinopoulos and Daniel Young, *Frappé Nation* (Potamos, 2006).

bitterness. The most popular Dutch and Belgian roasts were also medium-dark, reflecting their ties to Robusta-producing former colonies. Italy developed regional preferences, with an increasing emphasis on Robusta the further south one travelled. While this initially reflected the coffee's cheapness, it became an enshrined consumer preference. Portugal developed its own espresso-style beverage—the *bica*—brewed with blends of Robusta from Angola, where production, principally from white-owned estates, reached 225,000 tonnes in the early 1970s until the outbreak of the independence wars disrupted the industry.

The coffee industry's evolution in Europe during the twentieth century led to distinctive 'national tastes', with markets adopting different forms of brewing technology, roasts, blends and consumption habits. Many 'traditional' coffee styles we associate with particular countries date back no further than the last century. Often these were only consolidated in the mass consumer societies that arose after the Second World War. Local and regional roasters lost out as small grocers and specialist coffee shops declined, and supermarkets took over. These stocked the highly visible branded products that shoppers recognized from expensive television advertising campaigns on behalf of dominant 'national' roasters. Consumed in a similar fashion across classes, drinking coffee became an everyday expression of 'national identity'.

Germany

Germany became Europe's largest coffee market following its 1871 unification. By the 1900s coffee was drunk daily across all regions and classes, providing families with a warm accompaniment to morning and evening meals of bread or potatoes. Surrogate products satisfied a significant portion of this demand: in 1914, 160 million kilograms (350 million lb) of coffee substitutes were consumed along with 180 million kilograms (400 million lb) of coffee.[1]

Consumption rituals developed, such as the *Kaffeeklatsch*—women gathering for an afternoon chat with coffee and cake. Coffee stalls flourished in large cities, serving workers going to, or on a break from, the factory or office. At Sunday family gatherings, the 'best' coffee—made with real beans, not ersatz substitutes—would be produced, while summer afternoons might culminate with a stroll to the local park's coffee gardens. In some of these, hot water was sold, allowing less well-off patrons to brew up coffee they had brought with them.

Hamburg became Europe's leading coffee port, with around 90 per cent of imports now coming from Latin America, many shipped through the networks established by German émigré enterprises.[2] The coffee exchange, founded

---

[1] Julia Rischbieter, '(Trans)National Consumer Cultures: Coffee as a Colonial Product in the German Kaiserreich', in *Hybrid Cultures—Nervous States*, ed. U. Lindner et al. (Amsterdam, 2010), pp. 109–10.

[2] Dorothee Wierling, 'Coffee Worlds', *German Historical Institute London Bulletin*, XXXCI/2 (November 2014), pp. 24–48.

in 1887, counted some twohundred member organizations of importers, brokers and merchants by the start of the twentieth century. Consignments were graded and sorted by women warehouse workers, with dispatch samples placed in special coffee 'postboxes' in the port. Rail networks carried the coffee to the numerous wholesalers and grocers around the country.

In the 1880s van Guelpen, Lensing and von Gimborn, the Rhineland-based engineering firm that later became Probat, started producing drum roasters. These were used to set up wholesale coffee-roasting businesses supplying predominantly loose beans to local grocery stores. It was only later that roasters established their own branded identities, mostly confined to regional markets.

The major exceptions were brands linked to larger direct distribution operations, such as Kaiser's Kaffee, the ownbrand food store chain, which had 1,420 stores on the eve of the First World War. Eduscho, the Bremen-based mail-order house, became the largest roaster in Germany during the 1930s. In the immediate post-war era Tchibo, founded in Hamburg as a mail-order operator in 1949, transformed itself into West Germany's leading roaster, with a chain of small outlets during the 1950s and '60s. Here coffee could be sampled in-store and beans purchased for home. In the 1970s Tchibo extended into in-store operations within bakeries and supermarkets. In 1997 Tchibo and Eduscho merged into one company.

The German preference was increasingly for filter coffee.

In 1908 Dresden housewife Melitta Bentz patented a new preparation system using paper filters, set into a brass filter pot with punched holes. Supposedly she developed this by experimenting with her son's school blotting paper. Previous filter systems had relied on using cloths that were washed and re-used. Now housewives could clear up by throwing out the filter paper and grounds. Her husband established the company bearing her name, which was immediately successful, cementing its hold in the 1930s when the now-familiar coneshaped filters and papers were introduced.

Scandinavia

The Nordic countries also developed a strong regional coffee culture. This arose from the combination of coffee's functional benefits as a warming beverage against the cold, and the temperance movement, closely linked to the church, which promoted coffee as an alternative to alcohol. Coffee consumption per capita in Denmark and Sweden already exceeded the United States in the 1930s. By the 1950s Finland had the highest coffee consumption per capita in the world, closely followed by Norway.

Sami reindeer herders routinely drank twelve cups of coffee a day in the 1950s. The men warmed themselves with coffee before leaving in the morning. When the women heard the dogs barking announcing their return, they prepared fresh coffee. Coffee drinking was central to their hospitality ritual. This demanded visitors be offered, and accept, a minimum of two cups of coffee per visit, which could push up daily consumption to twenty cups. [1]

Urban communities also developed routines and rituals organized around coffee. *Fika*—a time to share coffee and cake with family, friends or one's work colleagues—evolved in Sweden during the late nineteenth century. Its centrality in the country's culture is such that today refugees are inducted into the practice. The Danish language acquired specific terms to describe the coffee prepared for a woman giving birth and the attendant midwife. Finnish labour legislation officially established coffee breaks during the working day. [2]

The Scandinavian preference is for light-roasted coffee. Industrial roasters perpetuate this style to reflect national distinctiveness. Paulig, the major Finnish roaster, placed an image of a young woman in national costume pouring coffee from a kettle on its branded products in the 1920s. Since the 1950s, Paulig has selected a young woman to be the Paula girl, who promotes the brand through public appearances.

Italy

Italy evolved a distinctive European coffee culture, due

[1]   L. Whitaker, 'Coffee Drinking and Visiting Ceremonial Among the Karesuando Lapps', *Svenska landsmål och svenstkt folkiv* (1970), pp. 36–40.
[2]   Dannie Kjeldgaard and Jacob Ostberg, 'Coffee Grounds and the Global Cup: Global Consumer Culture in Scandinavia', *Consumption, Markets and Culture*, X/2 (2007), pp. 175–87.

to its development of espresso brewing. [1] The spread of upmarket cocktail bars where drinks were quickly prepared and passed to customers across the counter led to a demand from the hospitality industry for equally swift ways of serving coffee. Applying pressure to the brewing process speeded up the extraction time, enabling a fresh cup of coffee to be prepared 'expressly' for each customer. The first commercially produced machine was the La Pavoni Ideale manufactured in Milan in 1905. It incorporated a boiler from which steam was drawn to drive hot water down through the coffee clamped onto a delivery outlet ('group head'). As the pressures were relatively low (1.5 to 2 bar), the preparation process still took around a minute, and produced a concentrated filter-coffee taste. These large, highly decorative machines sat on the bar counters of many top European hotels. The Fascist regime's suspicion of coffee as a 'foreign luxury' meant ordinary Italians were more familiar with coffee substitutes.

This changed after 1948 when Achille Gaggia produced a new espresso machine that utilized a lever connected to a spring-loaded piston to blast water through the coffee. It achieved higher pressures (around 9 bars), the delivery speed was much quicker (around 25 seconds) and the resultant extract was topped with a mousse or *crema* of essential oils. Subsequently manufacturers, notably Faema, introduced semiautomatic machines, replacing the piston with an

---

[1] Jonathan Morris, 'Making Italian Espresso, Making Espresso Italian', *Food and History*, VIII/2 (2010), pp. 155–83.

electric pump. A coffee from the bar now looked, and tasted, different from anything prepared at home. The same was true for cappuccino—originally meaning coffee with milk, but now used exclusively for espresso with steamed milk, and only available at the bar.

The 1950s and '60s saw the emergence of modern Italian coffee culture. Industrialization and urbanization led to an increase of neighbourhood coffee bars serving the small workshops and housing estates generated by migration from countryside to city. The speed of coffee preparation and drinking made bars ideal places for grabbing a cappuccino before work and for quick breaks during the day. Drinking coffee on one's feet became standard, not least because of a 1911 law allowing councils to impose a maximum price for 'a cup of coffee without service'—that is, standing at the counter. Set low to curb inflation, this made the coffee bar sector unattractive to corporate chains.

A key advantage of the espresso process is that it intensifies flavours so cheaper commodity beans can form significant portions of the blend. In the post-war era Brazil sold off stocks of low-grade Santos to Italy, where roasters also turned to Robusta, which had the added advantage of producing a thicker, visually appealing *crema*.

Domestic consumption doubled between 1955 and 1970. The aluminium, eight-sided stovetop brewer known as the 'Moka Express', manufactured by Bialetti, became standard equipment in Italian kitchens. It functions as a percolator: water heated in the lower chamber is forced up through the coffee by steam pressure to collect in the serving section at the top. It was advertised as producing coffee 'just like that at the bar', even though no *crema* is created.

In the 1960s the Lavazza company from Piedmont became the first coffee producer with a nationwide presence. Its success was due to an innovative television-advertising campaign using animated cartoon characters, combined with an extensive distribution system that penetrated the many neighbourhood stores throughout the country. In 1995, a hundred years after Luigi Lavazza opened a grocery roasting coffee in Turin, the company had a 45 per cent share of the 'at-home' Italian market.

Central Europe

The Viennese coffee house reached the peak of its fame in the early twentieth century. This success coincided with the democratization of culture and consumption that characterized the European *fin de siècle*. By 1902 there were around 1,100 cafés in the city, appealing to a broad middle-class clientele, along with over 4,000 working-class taverns.[1]

The public accessibility of the coffee houses was a key part of their appeal to groups such as Vienna's Jewish population who still encountered prejudice within Austrian society. The Young Vienna literary circle, containing many

---

[1] Charlotte Ashby, Tag Gronberg and Simon Shaw-Miller, eds, *The Viennese Café and Fin-de-siècle Culture* (London, 2013).

Jewish writers, met at the Café Griensteidl. Socialist thinkers including Leon Trotsky frequented the Café Central. These groups would occupy a *Stammtisch*, a table reserved for regulars who would drop in over the course of the day. The largely male waiting staff, presided over by a major domo known as *Herr Ober*, meant there was little likelihood of coffee houses being mistaken for houses of ill repute. Female guests were welcome, but dark interiors and the masculine atmosphere meant many women preferred to meet each other for coffee at the cake shop or *Café-Konditorei*.

The coffee house phenomenon extended throughout the Austro-Hungarian Empire. In Budapest, there were around five hundred coffee houses operating in the early 1930s, the most beautiful surely was the New York, established in 1894. Trieste, the empire's outlet on the Adriatic, became one of Europe's leading coffee ports, a status it maintained after being transferred to Italy after the First World War. This was how Ferenc Illy, born into a Hungarian family in what is now Timis, oara in Romania, came to found one of Italy's leading coffee-roasting companies in 1933, having stayed in Trieste after serving in the Austro-Hungarian army during the war.

Julius Meinl, the proprietor of a Viennese colonial goods shop established in 1862, set up a roasting company that developed into Central Europe's largest coffee supplier under the leadership of his son Julius Meinl II. By 1928 it was operating 353 grocery stores in Austria, Hungary, Czechoslovakia, Yugoslavia, Poland and Romania. In 1938 Julius Meinl III, a prominent anti-Nazi married to a Jew, moved the family to London, returning after the war to rebuild the company's fortunes.

The Austrian coffee house developed an extensive menu of coffee beverages. In addition to the black, the brown, the gold and the *melange* (mixture), whose names capture the relative proportions of milk and coffee, more esoteric offerings included the one-horse carriage (a large amount of whipped cream on top of a black coffee in a glass) and the Sperber-Turk—a double-sized Turkish coffee boiled with a cube of sugar, first consumed by a famous lawyer. [1] From the 1950s, however, espresso machines were swiftly adopted in Viennese coffee houses, so that the *melange* and the cappuccino became very similar.

Elsewhere in Central Europe, coffee culture had to coexist with communism, and vice versa. The 'Standard Coffee Blend' made available in countries such as Czechoslovakia was often of dubious origin and content: in 1977 East Germany reacted to a currency crisis by launching Kaffee-MIX—ground coffee combined with roasted peas, rye, barley and sugar beet. In Hungary, neighbourhood *eszpresso* bars were concessions to the ingrained coffee culture, but it was only after 1989 that the New York and other Budapest coffee houses were restored to their original splendour. Coffee consumption in East and Central Europe has risen dramatically since the fall of the Berlin wall.

---

[1] Harold B. Segel, *The Vienna Coffeehouse Wits, 1890–1938* (West Lafayette, IN, 1993), p. 11.

Japan

During the latter twentieth century, consumer markets outside Europe and North America assumed a significance within the global coffee trade. Mostly fuelled by convenience products, this resulted in the opening of long-established elite coffee cultures to the broader population.

Japan, today the world's third-largest importing country, exemplifies this. Coffee was first introduced to Japan by the Dutch East India Company in the late seventeenth century. It was confined to Dejima, the artificial island off Nagasaki, through which foreign trade was conducted during the Tokugawa shogunate, which maintained Japan as a 'closed country'. The only Japanese with a taste for coffee during this period were the island prostitutes who prized it for keeping them awake in order to prevent their clients from departing without paying. [1]

Coffee entered Japanese society in the later nineteenth century, after the Meiji restoration and American pressure to open the country for international trade. In 1888 Tei Ei-kei established the Kahiichakan, a coffee house modelled on those he experienced in New York and London. It was modelled on elite clubs: leather armchairs, carpets, newspapers, billiard tables and well-stocked writing desks. Sadly, selling access to all this for a single cup of coffee was

---

[1] Merry White, *Coffee Life in Japan* (Berkeley, CA, 2012).

not a viable business. Tei Ei-kei went bankrupt, dying in penury.

More commercial propositions were developed. The Café Paulista, a chain of *kissaten* (coffee shops with waiter service), were opened in the 1900s by Mizuno Ryu, who had been employed by Brazilian coffee plantations as part of a sponsorship programme to bring in Japanese workers as *colonos* after the Italian schemes ended. Japanese émigrés also worked on Hawaiian coffee farms during this era. As in Europe, the development of mass coffee culture in Japan was held back by the emphasis on autarchy in the interwar era.

Coffee import restrictions were only lifted in 1960. Japan imported 250,000 bags the following year. By 1990 the figure was 5.33 million bags. *Kissaten* started to spread through Japanese society from the mid-1960s. Proprietors were attracted because of relatively low entry costs, while customer numbers were fuelled by the country's economic boom. By 1970 there were 50,000 *kissaten*, peaking at 160,000 in 1982. Thereafter a division developed between older-style *kissaten* and Westernstyle, self-service café formats appealing to younger consumers. These included the Doutor coffee chain, which opened its first branch in 1980 and today has over nine hundred outlets.

Many *kissaten* developed their own preparation techniques and rituals, using special net filters or syphon equipment (originally introduced by the Dutch). This created opportunities for companies like Hario, a glassware maker, to produce highspecification, heat-retaining brewing equipment. The beans' origins also came to be appreciated, with Tanzanian Kilimanjaro gaining popularity due to the mountain's resemblance to the sacred Mount Fuji.

Mass-market development was driven by new convenience formats, notably ready-to-drink (RTD) beverages in cans. These were first produced by the Ueshima Coffee Company, which launched 'UCC Coffee with Milk' in 1969. The introduction of vending machines serving these hot and cold in 1973 created a substantial on-the-go 'industrial' coffee market. It is indicative of the popularity of sweetened beverages that it was nearly twenty years before UCC added a black sugarless coffee to its range.

Nonetheless it was soluble coffee that sparked the 'athome' market growth, accounting for five out of 8.5 cups consumed weekly in 1983, enabling Nescafé to become a market leader. [1]

International Coffee Agreement

The politics of the global coffee trade reflected shifts in economic power within both the consumer and the producer segments. The emergence of large-scale national roasters exercising significant buying power was intensified when American grocery multinationals began buying brands to access overseas markets. In the 1970s General

---

[1]   All Japan Coffee Association, 'Coffee Market in Japan', pdf document, www.coffee.ajca.or.jp/English, accessed 21 August 2017.

Foods acquired the Swedish roaster Gevalia and Sara Lee bought Douwe Egberts, based in the Netherlands. Phillip Morris added Jacobs Suchard to its roster in 1990. Nestlé, meanwhile, reversed the trend by acquiring U.S. brands such as Hills Brothers and Chase & Sanborne in the 1980s.

The rise in Robusta production fundamentally altered the supply side of the international coffee market. Prices fell after 1954 due to the world supply increasing, leading the Latin American producers to start limiting their exports in 1957. Their intervention failed because, as Robusta was priced lower than Arabica, new producers were unconcerned by Brazilian threats to dump coffee on the market. Withholding supplies simply incentivized buyers to switch to Robusta, leaving Brazil holding stocks equivalent to annual world consumption by 1959.

The Latin American countries began lobbying the United States for a global agreement introducing controls on imports as well as exports. They exploited political fears raised by the 1959 Cuban Revolution, with a Colombian senator urging, 'Pay us good prices for our coffee or— God help us all—the masses will become one great Marxist revolutionary army that will sweep us all into the sea.' [1] Given their reliance on Brazil and Colombia, the major U.S. roasters found it prudent to support an agreement while the Cuban Missile Crisis convinced Congress to ratify it. European consumer countries adhered so that their

---

[1] Gregory Dicum and Nina Luttinger, *The Coffee Book* (New York, 1999), p. 86.

remaining colonies and newly independent producer states could achieve economic security.

The 1962 International Coffee Agreement (ICA) was signed by 44 exporting members and 26 importing ones. Its stated aim was:

> to achieve a reasonable balance between supply and demand on a basis which will assure adequate supplies of coffee to consumers and markets for coffee to producers at equitable prices and which will bring about long-term equilibrium between production and consumption. [1]

The agreement then established the International Coffee Organization (ICO), headquartered in London.

The ICO Council formed the supreme body for implementation. Proposals had to gain 70 per cent of the votes from producer and consumer members. Votes were assigned in proportion to the volume of members' exports or imports. Brazil held 346 of the 1,000 producer votes, and the USA four hundred of the 1,000 consumer ones. Target price bands for four forms of coffee were established: Colombian Milds, Other Milds, Brazilian Naturals and Robusta. Members were assigned quotas for each export type. When prices rose above the band, as they did following the devastating Brazilian frost of 1975, export quotas were relaxed to bring them down; if they fell below the band, they were tightened

---

[1]   Richard Bilder, 'The International Coffee Agreement', *Law and Contemporary Problems*, XXVIII/2 (1963), p. 378.

to raise it back up. The quota regime remained in operation from 1962 to 1989.

The balance of power in the global coffee chain shifted towards producer states, more specifically the quasi-state agencies representing them in the ICO. When consumer members were reluctant to enforce quotas in the early 1970s following the collapse of the Bretton Woods exchange rate system and the oil price shock, the largest producer agencies such as the IBC, FNCC and Ivorian *Caisse* collaborated by establishing joint entities to buy and sell coffee on the world markets. Using their specialist 'insider' knowledge of stocks and harvest forecasts, they successfully frustrated financial speculators efforts to manipulate futures prices.

The quota system remained in place throughout the 1980s, primarily for political reasons. Following Nicaragua's 1979 Sandinista revolution, the U.S. Reagan administration wanted to avoid further left-wing triumphs in the civil wars in El Salvador, Guatemala, Nicaragua and Colombia. These partially derived from the wealth distribution iniquities created within their coffee-growing sectors. In Guatemala, 1 per cent of the coffee farms produced 56 per cent of the crop. [1] As in El Salvador, the indigenous peasantry who worked the farms became the targets of violent ethnic repression by military rulers. Conversely, guerrillas demanded 'war taxes' from middle-class farm owners, on pain of their

[1] Steven Topik, John M. Talbot and Mario Samper, 'Globalization, Neoliberalism, and the Latin American Coffee Societies', *Latin American Perspectives*, XXXVII/2 (2010), p. 12.

buildings being burnt down and lands occupied. Under Sandinista rule, ENCAFE, the Nicaraguan coffee agency, returned only 10 per cent of the coffee export price to producers. [1]

The relatively secure returns delivered by the quota system compared to other commodities encouraged producers like the Philippines and Indonesia to increase their coffee output. An additional incentive was that the ICA established quotas balancing a 'demonstrated capacity' for production against 'historic' world-market share.

Most new production was Robusta, which was in demand for instant coffee products. Keen to exploit these new cheap supplies of coffee, importers developed 'coffee-cleaning' techniques, steaming the beans to moderate their bitterness. By 1976 Nestlé had established processing subsidiaries in 21 producer countries. [2] Some states set up their own enterprises: Ecuador planted and processed Robusta for export as coffee powder. Brazil introduced Conilon, a Robusta variety, in the state of Espirito Santo and developed a processing infrastructure. Soluble coffee has gained popularity among Brazilians, and today around 20 per cent of all Brazilian production is Robusta.

Some Central American state agencies took advantage of the stability to invest heavily into agricultural research to improve yields. So-called 'technification', including the introduction of dwarf cultivars capable of growing in full

---

[1] Mark Prendergast, *Uncommon Grounds* (New York, 2010), p. 317.

[2] John M. Talbot, *Grounds for Agreement* (Lanham, MD, 2004), p. 61.

sun, and the use of chemical fertilizers, resulted in producers dramatically increasing outputs: between the mid-1970s and early 1990s, Colombia's harvest rose by 54 per cent, Costa Rica's 89 per cent and Honduras's by 140 per cent.[1] Rather than destroy excess coffee production, exporters disposed of it cheaply into markets not covered by the ICA, such as the Soviet Bloc. In 1989, 40 per cent of Costa Rica's crop was sold at half price or less and some was used as barter payment for goods such as Czechoslovakian buses.

Roasters' desire to access new supplies often resulted in a 'tourist' coffee arriving into a quota country via a non-quota one. Some further sleight of hand at the receiving port could result in high-grade coffee designated for a non-member country being swapped with low-grade coffee intended for a member one, so all parties obtained coffee at below-quota prices. The phenomena persisted because members were reluctant to adjust quotas at their own expense to reflect shifting demand.

In September 1989, with Soviet and Sandinista regimes fading, the United States withdrew its support for the quota system, eventually withdrawing from the ICO in 1993. It was the only consumer member to do so, but without it there was no way to implement a regulatory system. Meanwhile the contrasting interests of producer nations left few with any appetite to continue. Some state agencies were disbanded, notably the Brazilian IBC. Today, the ICO continues

---

[1]   Ibid., pp. 77–81.

operating as an international information exchange, but lacks any global coffee supply chain governance role.

For all its dysfunctionality, the quota regime delivered relative stability. The monthly indicator price varied by 14.8 per cent during the last eight years of quotas; over the following eight years, variability was 37 per cent. Between 1984 and 1988 the average indicator price was $1.34/lb; between 1989 and 1993, as supplies flooded the market, that fell to $0.77.[1] Brazilian frosts halted the fall as supplies fell, but it was clear that the systemic instability of the global coffee trade had returned.

Vietnam

Vietnam exploited the ICA'S demise, fundamentally changing the world coffee trade in the twentieth century's last decade. It became the world's second-largest coffee producer in 1999, overtaking Colombia, having ranked only 22nd in 1988. The key to its success was growing Robusta, of which the country became the world's largest exporter.

Missionaries planted small amounts of Arabica in the 1850s, but coffee remained a minor crop during French colonial rule. By 1975, at the end of the long war between the communist North and American-backed South, only 60

---

[1] International Coffee Organization, *World Coffee Trade (1963–2013): A Review of the Markets, Challenges and Opportunities Facing the Sector* (London, 2014).

hectares (148 acres) of coffee remained. Following its victory, the Communist regime sought to stabilize areas formerly under southern control. Loyal North Vietnamese peasants were encouraged to migrate into the Central Highlands region where state farms and agricultural cooperatives were established, through a combination of nationalization of existing holdings and an aggressive deforestation programme. Here they were encouraged to grow coffee for export to Vietnam's Soviet Bloc allies.

It was only after the government economic reforms started in the 1980s that output began to soar. Transferring land into private hands during the 1990s meant that by 2000, 90 per cent of coffee production was undertaken by smallholders farming plots of less than 1 hectare (2½ acres). They were still strongly supported by the state, which continued to incentivize production through land subsidies, financial credits and technical assistance, such as access to fertilizers. This resulted in a remarkable average 24 per cent per annum growth in output between 1988 and 1999, and much higher average yields than competitor countries.

In 1995, Vinacafe, the state body responsible for the coffee industry, including development, marketing and export activity, was transformed into the Vietnamese National Coffee Corporation. It operates the remaining state farms as well as many processing, trading and service providers. It controls around 40 per cent of the country's exports, and operates one of the country's two instant coffee facilities (the other belongs to Nestlé).

The rapid expansion of Vietnamese production seemed designed to shore up the regime's political position in the wake of the Soviet Bloc's demise. It increased its export revenues while allowing peasants to 'enrich themselves' through direct market contact. The danger was that eventually the glut of supplies would drive down prices—as happened dramatically after 1998.

The Coffee Crisis

In 1998 the ico's composite indicator price for coffee was 109 U.S. cents per pound; by 2002 it had fallen to below 48 cents. Although the price rose thereafter, it was only in 2007 that it returned above $1 to 107 cents. This dramatic and prolonged price collapse had profound effects. Producers were plunged into poverty, while the industry's public image in consumer countries came under attack.

The problem was an ongoing excess of supply over demand. In 2001–2, 113 million bags of coffee were produced, an additional 40 million bags had accumulated, yet world consumption stood at 106 million bags. The ico executive director declared, 'At the origin of this coffee glut lies the rapid expansion of production in Vietnam and new plantations in Brazil.'[1]

[1] Néstor Osorio, 'The Global Coffee Crisis: A Threat to Sustainable Development', Submission to World Summit on Sustainable Development (Johannesburg, 2002).

The price fall consequences were felt differently across the sector. Where costs of production were low, technologies well developed and exchange rate movements favourable, such as in Brazil, it was still possible to make profits. Conversely, where coffee was used as a cash crop by subsistence farmers, as in most African, some Central American and many Asian countries, this reduced the money available to spend on medicines, education, food or servicing debts.

In Guatemala, the coffee labour force was halved. In Colombia, farmers ripped up their coffee trees, replacing them with coca plants for the drugs trade. Many Mexican growers gave up and attempted to illegally enter the U.S., often perishing in the attempt. Political conflicts intensified, with peasants in Chiapas, the centre of Mexican coffee production, supporting the Zapatista guerrilla movement's rebellion against the government.

Even in Vietnam, some farmers were forced to sell possessions to satisfy debt collectors. Poverty levels in the Central Highlands reached 50 per cent, with 30 per cent of the population suffering from hunger and malnutrition. Robusta's price fell from 83 cents in 1998 to just 28 cents in 2001. This had huge repercussions for countries like Uganda that were heavily dependent on exporting coffee.

Vietnamese output continued to expand, however, as peasants sought to produce their way out of the crisis. In 1990–91 the country produced 1.3 million bags, in 2000–2001 14.8 million bags, and by 2015–16 an astounding 28.7

million bags, more than the entire continent of Africa. Other Asian countries followed Vietnam's lead. The Southeast Asian countries Myanmar, Laos and Thailand developed significant coffee industries. Indonesia and India joined the ranks of the top six producers. In all these cases, over 90 per cent of the output is Robusta.

Coffee prices accelerated after 2010, when a virulent attack of coffee rust started spreading throughout Latin America. This has forced a rebalancing of supply and demand, particularly for high-quality Arabica, resulting in coffee's composite price remaining at over 120 cents per pound throughout the decade. It should not be forgotten that some of this new- found stability is at the expense of those forced out of farming, whether by falling prices, drought or disease.

Coffee's price volatility following the collapse of the quota system in 1989 is proof to many of the dangers of deregulation. Even the rust epidemic has been blamed on the demise of para-state agencies that coordinated national research and responses to crop disease. [1] Yet though the quota regime regulated coffee's flow onto the market, it favoured established producers, while state agencies frequently failed to pass back profits made from coffee to farmers.

The paradox that the so-called 'latte revolution', characterized by the rapid growth of coffee shops charging

[1] Stuart McCook and John Vandermeer, 'The Big Rust and the Red Queen', *Phytopathology Review*, CV (2015), pp. 1164–73.

premium prices, coincided with the coffee crisis, provoked criticism of consuming 'poverty in your coffee cup'. [1] Others, though, saw this new phenomenon as an opportunity to recast coffee as a 'specialty beverage', facilitating its de-commodification, and the generation of greater revenues throughout the value chain.

---

[1] Oxfam, *Mugged: Poverty in Your Coffee Cup* (Oxford, 2002).

Chapter 6
A Specialty Beverage

The repositioning of coffee as a specialty beverage at the end of the twentieth century has had profound effects upon the global coffee industry. What began as a protest by independent roasters in the U.S. against commodification and industry concentration spawned the spread of international coffee shop chains, the hipster 'third wave' movement, the development of the coffee capsule, and a set of fierce debates about ethical coffee consumption. Arguably, the role of specialty in stimulating consumption in non-traditional markets has laid the foundations for a new era in coffee history.

The Birth of Specialty

In the United States, the four leading roasters' market share rose from 46 per cent in 1958 to 69 per cent in 1978. By 2000, the 'big three'—Procter & Gamble, Kraft, and Sara Lee—controlled over 80 per cent of the retail market. They competed on price: blend contents were cheapened, and some brands advertised that lower quantities of their product could be used to deliver the same brew strength.

These tactics failed to reverse the steady decline in U.S. coffee consumption per capita, which dropped from around 7.25 kilograms (16 lb) in 1960 to 2.7 kilograms (6 lb) by 1995, despite the success of the automatic Mr Coffee machine at converting Americans (especially men) to drip brewing in the 1970s. By contrast, consumption of caffeinated soft drinks boomed, for reasons as diverse as the spread of central heating, the rise of fast food outlets and the appeal of youthful advertising.

Independent roasters saw their numbers fall from around 1,500 in 1945 to 162 in 1972. To survive they evolved an alternative business strategy. Rather than price, they would compete on quality, enabling them to increase profit margins on their beans. Their approach suited a consumer economy in which different social groups had started using their purchases to convey messages about their lifestyles, values and tastes. These might include demonstrating sophistication or wealth; adherence to 'alternative', anti- corporate values; or a preference for 'authentic' artisan goods.

Coffee was an important ingredient of 1960s American counterculture, whose spiritual home lay in San Francisco. Hippies hung out at North Beach espresso bars run by Italian immigrants, and purchased their beans from Alfred Peet's store in Berkeley. Peet, a Dutchman, roasted his coffee considerably darker, and brewed it much stronger, than a regular 'cup of Joe'. Despite the proprietor's barely disguised disdain for many of his customers, Peet's became a mecca for those keen to experience 'European' coffee.

The first person to use the term 'specialty coffee' was Erna Knutsen. In the mid-1970s she convinced the San Francisco coffee importers where she had started as a secretary to allow her to try selling small lots of quality coffees. She found a niche supplying a new generation of independent roasters, many of whom had 'dropped out' from conventional career paths.

In 1982 a group of roasters founded the Specialty Coffee Association of America (SCAA), defining 'specialty' as delivering a distinctive taste in the cup. Their product ranges included high-grade export coffees such as Kenyan AA, alongside blends and flavoured coffees with names such as 'Swiss Mocha Almond'—all unlikely to be classed as specialty today. These were sold in gourmet delicatessens popular with 'yuppies'—the young urban professionals whose rising purchasing power underpinned the 1980s foodie revolutions.

Specialty coffee took off once emphasis switched from selling beans to serving beverages. Seattle was at the centre: in 1980 the first coffee carts incorporating espresso machines appeared in the city; by 1990 there were over two hundred carts positioned close to monorail stations, ferry terminals and major stores. Workers, it turned out, preferred to pay for takeaway specialty-style beverages than consume the free coffee available in their offices. Today only one or two carts remain, the others swept away by the coffee shop revolution that the city spread to the world.

The Origins of Starbucks

Starbucks was set up by three college friends in 1971. It primarily sold beans supplied by Alfred Peet, whose dark roasting style they subsequently adopted. Howard Schultz, a salesman for a Brooklyn company which was one of their equipment suppliers, visited in 1982 and convinced the founders to hire him as sales and marketing director. In 1983 Schultz visited Milan, where he

> found the inspiration and vision that have driven my own life, and the course of Starbucks . . . If we could re-create in America the authentic Italian coffee bar culture...Starbucks could be a great *experience*, and not just a great retail store. [1]

He failed to convince the Starbucks owners of his case, however, and left to open a coffee shop called *Il Giornale* in 1986. He chose the name believing it meant 'daily', in reference to the frequency with which Italians visit their local bar. In fact, it means 'newspaper'.

This was not the only element of Schultz's vision that did not translate. Customers did not want to stand sipping

---

[1] Howard Schultz and Dori Jones Lang, *Pour Your Heart Into It* (New York, 1997), p. 52.

coffee at the counter, but to sit at a table and chat. They preferred paper cups to porcelain ones, so they could take their drinks back to work. Opera in the background and bow-tie-wearing baristas did not fit with the informal Pacific Northwest vibe.

Once Schultz adjusted his offer to create an 'Italianstyle' experience that met American customer needs, he started to have success. In 1987 he transferred this format into Starbucks, which he bought when the last of the original founders left for San Francisco to take over Peet's.

Coffee Shop Format

The coffee shop format combines two elements: the coffee and the environment. The former pays for the latter.

Italian-style coffees proved perfect for introducing American consumers to specialty coffee, as the distinctive bite of the espresso could still be discerned through the sweetness of the milk. Caffè latte was the most popular, as steamed, rather than frothed, milk produces greater density and sweetness than in a cappuccino. The addition of flavoured syrups allowed shops to develop bespoke ranges and offer seasonal beverages such as eggnog latte. Approachability proved more important than authenticity: a Starbucks standard tall cappuccino is twice the size of an Italian one—further augmenting the sweetness.

By 1994 espresso-based beverages were outselling

brewed coffees in U.S. specialty stores. The theatre of the barista 'hand-crafting' the beverage—grinding fresh beans, pulling a shot from the machine, foaming and pouring the milk, topping with cinnamon, chocolate and/or sprinkles—all rendered visible the value added during the process. Consequently, consumers were prepared to pay a high price for a premium product they could not make at home.

The high margins incorporated in the price paid for a comfortable environment in which the coffee could be enjoyed. Sofas, music, newspapers and clean toilets with babychanging facilities all helped create a 'twenty-minute business'. Coffee is the rental charge for using the facilities provided by the shop. It feels democratic, as customers are served in order of arrival at the counter; and inclusive, as the focus on coffee rather than alcohol renders it a 'safe' space for women, children and non-drinkers.

Schultz trumpeted Starbucks as an exemplar of a 'third place' between work and home in which—as the sociologist Ray Oldenburg describes it—informal contacts between unrelated people create a sense of community.[1] Behavioural studies, however, find little evidence of conversations being initiated between strangers: the attraction of the coffee shop lies in being surrounded by people without having to engage with them. The continuing advances of digital technologies—the laptop computer, the mobile phone, the wireless Internet connection—allow individuals to continue

---

[1] Ray Oldenburg, *The Great Good Place* (New York, 1989).

working, or engage in social media conversation, while 'consuming' the coffee shop ambience.

The Hegemonic Chain

Schultz proved adept at raising capital for expansion. After an initial public offering of stock in the company in 1992, Starbucks concentrated on acquiring high footfall locations, frequently close to each other on the same street. This had the effect of growing overall trade, because people would not deviate far from their daily routines to get a coffee. Once converted to specialty coffee, however, they were disposed to drink it wherever they found it, boosting not just Starbucks' trade, but that of the whole sector.

Branding was critical to maintaining Starbucks' premium position. Guaranteeing customers the same experience whichever outlet they visited required staff to follow a customer service script and brew the same beverages consistently, hence Swiss super-automatic push-button espresso machines replaced the traditional Italian equipment in 1999. Celebrities were paid to be 'found' and photographed sipping coffee from branded takeaway cups. Starbucks maintained itself as the hegemonic brand within the coffee shop sector, and was so dominant that it effectively defined what consumers understood the coffee shop concept to mean.

Specialty Coffee Shops and Starbucks outlets in the USA[1]

Year	Specialty Coffee Shops	Starbucks Outlets	Proportion (%)
1989	585	46	7.9
1994	3,600	425	11.8
2000	12,600	2,776	22.0
2013	29,308	11,962	40.8

In 2016, as many Americans reported drinking a 'gourmet coffee' or specialty coffee as a regular one the previous day.[2] This was driven by a threefold rise in the consumption of espresso-based beverages since 2008, which in turn reflected their adoption by everyday fast food chains and corner shops such as Dunkin' Donuts and McDonald's. Caffè latte is now as much the American coffee style as a 'cup of Joe'.

Internationalization

Internationalization was the other key strand to the Starbucks strategy—on 1 January 2017 there were 25,734 stores operating in 75 countries. The first opened in Japan and Singapore in 1996, and the programme was quickly extended into other Southeast Asian 'tiger economies'.

---

[1] Data available at www.sca.org and www.statista.com.

[2] 'What Are We Drinking? Understanding Coffee Consumption Trends', www.nationalcoffeeblog.org, 2016.

Coffee shop culture was warmly embraced by younger members of the middle classes, keen to emulate American trends by adopting them as hangouts for socializing and studying.

In Europe, the coffee shop concept often travelled in advance of Starbucks, as imitators and émigrés adapted it to meet local tastes. Costa Coffee, a London-based coffee roaster supplying the Anglo-Italian café trade, started opening espresso bars in the late 1980s. In 1995 it was purchased by the brewing and leisure conglomerate Whitbread, which correctly foresaw that coffee shops would replace pubs as social hubs in Britain. From 41 outlets in 1995 Costa reached 2,100 in 2017, making it the largest operator. Like all chains in the uk, it employs largely foreign baristas: young people taking advantage of then current European Union laws on freedom of movement. [1]

Branded chains drove uk coffee shop growth, but since the 2010s the fastest-developing sector has been among department store and garden centre chains, reflecting the extent to which Italian-style coffee has become a mainstream British beverage. Pubs, whose numbers continue to decline, are now serving coffee during the day as a survival strategy.

Coffee shop chains have spread throughout the Continent, but their progress has been contingent on the character of local coffee cultures. In Germany, Vanessa Kullmann set up the first chain, Balzac Coffee, in 1998, after

---

[1]   Jonathan Morris, 'Why Espresso? Explaining Changes in European Coffee Preferences', *European Review of History*, XX/5 (2013), pp. 881–901.

experiencing coffee shops as a fashion buyer in New York. Premium espresso beverages now constitute around 50 per cent of the German out-of-home market, but the majority are served in bakery chains. In France, where espresso-style beverages were well established, chains appeared in the 1990s, yet only really started to gain momentum as quick-service alternatives to bistros following the economic slump of 2008. The same occurred in Greece.

Italy, as the originator of espresso, has been a major beneficiary of the specialty revolution. Exports of roasted coffee rose from 12 million kg in 1988 to over 171 million kilograms in 2015. Italian companies control 70 per cent of the global market for commercial espresso machines, routinely exporting over 90 per cent of their output. Groups such as Illy and Segafredo have established branded chains throughout the world using licensing and franchising options. In Italy itself, however, no coffee shop chains have emerged, as it would be impossible to charge a premium price for espresso. Starbucks is due to open in Milan in 2018, some 35 years after Schultz's revelatory visit, but using a format designed to highlight its 'third wave' credentials.

The Third Wave

The term 'third wave coffee' was first used by Timothy Castle in 2000, and popularized by Trish Rothgeb, an

American roaster, in an influential article in 2003. [1] The first wave of mass-market roasters, she wrote, had 'made bad coffee commonplace'. The original specialty operators 'started destination shops with small roasting operations . . . serving espresso', but their format was eclipsed by second wave giants such as Starbucks who 'want to automate or homogenize specialty coffee'. The third wave would pursue a 'no rules' approach to crafting outstanding coffee.

Barista competitions are at the centre of third wave culture. The first World Barista Championships were held in Monaco in 2000. Competitors prepare a set of four espressos, cappuccinos and 'signature drinks' within fifteen minutes and are judged on technical and presentational skills, as well as the sensory qualities of their beverages. Equipment makers vie to have their machines classified as meeting competition standards. Roasters train baristas full time to compete using specially sourced blends. Winners gain celebrity status that bring high-paying contracts for consultancy and endorsements.

Third wave baristas experiment with the established parameters for espresso preparation and taste profiles, breaking away from Italian traditions. New beverages appeared as a result of those experimentations, such as the flat white, made using concentrated shots of espresso topped

---

[1]   Timothy J. Castle and Christopher M. Lee, 'The Coming Third Wave of Coffee Shops', *Tea and Coffee Asia* (December 1999–February 2000), p. 14; Trish Rothgeb Skeie, 'Norway and Coffee', *The Flamekeeper: Newsletter of the Roasters Guild*, SCAA (Spring 2003).

with velvety microfoamed milk and finished with latte art—all demanding high technical skills from the barista. The flat white was brought to London in 2007 by baristas from Australia and by 2010 had crossed into the mainstream chains, later crossing the Atlantic.

Third wave coffee shops often operate on a shoestring, their owners inspired more by passion than profitability. Stripped-back interiors and basic seating highlight the hightech machinery on the counter into which all the investment has been poured. This aggressively non-corporate ambience recurs so often that it has become the third wave's own brand image.

Third wave roasters source single-origin coffees from the same geographical district, preferably traceable back to a single farm or a producer cooperative. They take an artisan approach—roasting in small batches and adjusting profiles to achieve the best results from each lot. The roasts are usually light—designed to bring out the taste profiles of the beans, rather than draw attention to the roast's character itself.

In 1999 leading U.S. specialty buyers began organizing Cup of Excellence competitions in producer countries. Farmers submit their coffees to be assessed by an international jury of cuppers, with lots from the winners being auctioned online for astronomical prices. It is indicative of the global spread of specialty coffee that purchasers of the winning lots in the 2016 auctions came from Japan, Korea, Taiwan, Bulgaria, Australia, the Netherlands and the USA.

The third wave has increasingly shifted attention away from espresso to other forms of coffee brewing best able to bring out the subtleties of single-origin specialty grades. Equipment developed in Japan—such as the Hario V60 filter and the syphon brewer—has become common in third wave coffee shops. The Chemex, first manufactured during the 1940s, has acquired popularity, as its enhanced paper filters produce an exceptionally clean coffee resulting in beverages that can seem closer to tea.

The third wave can best be described as a form of transnational 'subculture', with its own mix of philosophies, iconic brands, fanzine-style publications and key influencers. The Internet has made this possible, enabling micro-roasters to find customers around the country—so-called 'prosumers'—to discuss the best ways to customize their machinery, and connoisseurs to read the latest coffee reviews online. These communities come together at coffee festivals such as that held in London since 2011.

Single-portion Coffee

The specialty revolution created a desire to prepare similar beverages in the home. Machines using 'single portion' coffee capsules have delivered this. Portions of ground coffee are sealed into aluminium capsules to preserve freshness. When placed into operating machines, the top of the capsule is punctured by pins and hot water is injected into it, causing the capsule to rupture under pressure as the coffee is delivered. Such systems combine convenience with cleanliness, but the capsules can only be recycled using specialized equipment, requiring consumers to make the commitment to collect and return them, rather than placing them in their household waste.

Nespresso, established by Nestlé in 1986, pioneered this technique for delivering espresso-style beverages, and remains the sector's global leader. In the U.S. market, the Keurig K-Cup system, introduced by Green Mountain Coffee Roasters in 1998, dominates the market for replicating American-style drip-brewed coffees.

Nespresso was developed for hospitality operators such as small restaurants, airlines and train companies, meeting their need for machines that did not require trained baristas or large amounts of space. Keurig was targeted to the office and hotel bedroom market, as providing pods prevented mess and waste. It soon became apparent, however, that these same advantages made the systems attractive to home users.

Nespresso positions its products as offering an entrée into the gourmet coffee world. Alongside espresso blends to suit a variety of palates and preferences, it offered socalled *grand cru* and limited-edition coffees to customers enrolled in its members' club. In 2000 the company opened its first retail boutique in Paris; by the end of 2015 there were 467 in sixty countries, occupying prime locations in major cities chosen for their proximity to luxury brand outlets. Co-branding strategies, such as the introduction of machines designed by Porsche, has cemented Nespresso as an upmarket lifestyle product, endorsed by George Clooney, who has been the principal brand ambassador since 2005.

Between 2000 and 2010 Nespresso experienced annual growth rates of over 30 per cent per annum. The premium it commands for its products is such that in 2010 Switzerland became the world's largest exporter of roasted coffee by value, even though it lies only fifth by volume.

In 2012 the patents protecting Nespresso and Keurig's proprietary technologies expired. In the five years thereafter, the global market for single-portion coffee grew by at least 50 per cent. Coffee shop chains capitalized on their brands by launching home systems, and manufacturers attempted to undercut Nespresso and Keurig by producing compatible pods, while artisan operators explored the potential of pods for third wave coffee. By 2017 at least a third of coffee-drinking households in the U.S. and UK were making use of capsule machines.

Ethical Coffee

Specialty coffee played the leading role in the adoption of certification systems whose labels attest to the environmental or socioeconomic sustainability of the supply chain for a particular coffee. Environmental certifications include Organic, Bird Friendly and Rainforest Alliance, which promote sustainable farming techniques that encourage biodiversity.

The first social certification programme was developed by the Fairtrade movement. In 1988 Solidaridad, a Dutch religious organization, established the Max Havelaar label— named after the novel denouncing the colonial coffee trade in Java. It started purchasing from producer cooperatives, initially in Mexico, and marketing the coffee in Germany and the Netherlands. In 1989 UK charities including Oxfam followed suit, creating the Cafédirect brand, and selling it through church halls and charity shops. In 1997 Fairtrade International was established to unite the various national schemes.

Fairtrade remains the only certification system to guarantee producers a minimum price for their coffee. Its pricing structures reflect whether the coffee is Arabica or Robusta, natural or washed, organic or non-organic. In addition, the exporting cooperative receives a social premium to be invested in improving the living conditions of the coffee-farming community. Since 2011 a quarter of this premium must be invested in improving quality. Should the world market price exceed the Fairtrade price at the time of delivery, then the higher price applies.

The Fairtrade organizations do not themselves buy or sell coffee. Instead they grant permission and charge for products to be labelled as 'Fairtrade'. For this to happen, all the participants in the commodity chain must be certified to ensure adherence to Fairtrade standards. Critics of

Fairtrade argue that the price guarantee induces inertia among producers by protecting them from the market and the need to respond to it. Furthermore, the bulk of the price differential paid by the consumer remains within the developed world, either with the roaster or supporting the operational costs of the certification systems.

Analysing the impact of Fairtrade certification systems upon producers has revealed mixed results. The floor price provided a significant safety net during the mid-2000s coffee crisis and many communities benefitted from the reinvestment of the social premium. Since then the gap between the Fairtrade and market price for coffee has remained relatively narrow. Research in Latin America suggests that this differential is not always sufficient to offset the potential losses in farmers' incomes from requirements to pay pickers better wages, and the organic premium does not make up for the reduced yields resulting from conversion. [1] A 2017 study discovered that while Fairtrade producers in Asia earn sufficient income to support their households, those in Africa cannot because their holdings are too small to reap the benefits of the premium. [2]

The value of the Fairtrade label to roasters and operators is that it demonstrates their ethical convictions, while enabling them to charge a premium that covers the

---

[1] Daniel Jaffee, *Brewing Justice* (Berkeley, CA, 2007); Tina Beuchelt and Manfred Zeller, 'Profits and Poverty', *Ecological Economics*, LXX (2011), pp. 1316–24.

[2] True Price, *Assessing Coffee Farmer Income* (Amsterdam, 2017).

additional purchase costs. They became particularly sensitive to this during the coffee crisis years, when the paradox of premium-priced lattes and starving coffee farmers was regularly highlighted in the media. This contributed to Starbucks being targeted during the antiglobalization riots in Seattle in 1999, although it was high-volume commodity roasters who were paying the least to their suppliers.

There was, however, relatively little Fairtrade coffee available, because of the organization's insistence that this should be sourced through producer cooperatives. This excluded a priori coffee from large plantations, independent farmers and smallholders who were not attached to cooperatives, as well as a significant number of cooperatives which were also put off by the initial certification costs.

Alternative certification schemes evolved that could be accessed by these producers, but left traders and producers to determine the premium placed upon the label. The Common Code for the Coffee Community (known as 4Cs), developed by large roasters and producer states, introduced a baseline set of social, environmental and economic standards in 2007. The price of 4C producer certification is graduated according to output, making it within the reach of many smallholders, while its easily achievable standards make it attractive to multinational corporations sourcing from multiple suppliers.

In 2012 Fairtrade USA broke away from Fairtrade International to enable it to certify non-cooperatively organized producers. It argued that this extended protection to labourers and independent smallholders, while offering more consumers the choice to buy Fairtrade.

By 2013 around 40 per cent of coffee's global production was in accordance with some form of certification standard. [1] Enthusiasts have argued that this represents one of the greatest triumphs of imposing social responsibility on global capitalism. Critics say this is a triumph of public relations, enabling the coffee industry to simultaneously monetize consumers' ethical concerns while engaging in 'virtue signalling'.

Third wave roasters object to Fairtrade's lack of concern with quality and traceability. Buyers for Stumptown, Intelligentsia and Counter Culture, leading third wave coffee roasters in the USA, developed an alternative model of 'direct trade': identifying growers of potentially outstanding coffee, working with them to ensure quality and purchasing from them directly at prices that reflect this, way beyond those achievable under Fairtrade. Such partnerships can be transformative in their impact on growers, but are confined to farmers in locations where cultivating specialty coffee is possible.

Interventions linking Fairtrade, direct trade and other development programmes have done much to improve conditions in certain countries. Rwanda's coffee infrastructure, which centred on commodity production, was destroyed during the 1994 genocide. In 2002 a national coffee strategy was introduced, investing in the installation of

---

[1]   David Levy et al., 'The Political Dynamics of Sustainable Coffee', *Journal of Management Studies*, LIII/3 (May 2016), p. 375.

washing stations, part-financed by foreign aid programmes. Roasters have developed direct relationships with farmers and processors, investing in training and constructing cupping labs to test for quality. Fairtrade organizations have certified cooperatives that operate the washing stations.

Rwanda nearly doubled the value of its coffee exports between 2006 and 2012 because of the much higher prices enjoyed by fully washed coffees. Much of this additional revenue is returned to producers; and encounters between farmers of different ethnicities using the washing stations have contributed to lowering tensions. [1] Rwanda is now widely recognized as a specialty producer; in 2008, it became the first African country to host a Cup of Excellence competition.

Globalizing Consumption

The specialty movement's greatest impact upon the global structures of coffee may well be its role in encouraging consumption in emerging economies, especially in producer states. By repositioning coffee as an aspirational beverage, it has become popular among younger, Western-orientated consumers.

V. G. Siddhartha opened the first outlets of Café Coffee

[1] Karol C. Boudreaux, 'A Better Brew for Success: Economic Liberalization in Rwanda's Coffee Sector', in *Yes Africa Can: Success Stories from a Dynamic Continent* (World Bank, 2010), pp. 185–99.

Day, the Indian coffee shop chain in Bangalore, in 1996, tailoring them to a very specific demographic:

> We designed the place as one where students and youngsters would hang out. The internet was just coming in. We thought this would appeal to the software crowd in Bangalore, who had some international exposure. We started an internet service and gave the coffee free. Or we said, you buy a coffee you get half an hour internet service free . . . Seventy per cent of India is below the age of 35 as are 80 per cent of our customers . . . They want the same experience that they see in the movies or television or get through the internet. [1]

In 2015 Café Coffee Day was operating over 1,500 outlets in India. A critical part of both its business plan and customer proposition is growing its own coffee, controlling the process from crop to cup.

China is the market with most potential in Asia, by its size and current state of underdevelopment. Between 2004 and 2013 consumption grew by 16 per cent per annum, yet still only reached 83 grams per capita—enough for five or six cups of coffee *a year*, in what remains an overwhelmingly tea-drinking culture.

In the much smaller out-of-home sector, however, coffee accounts for 44 per cent of all sales. China already

---

[1]   Ashis Mishra, 'Business Model for Indian Retail Sector', *IIMB Management Review*, XXV (2013), pp. 165–6.

hosts the largest number of Starbucks outlets outside the USA, even though their clientele is largely confined to the wealthy urban middle class. Chinese consumers are particularly attracted to using their mobile phones to order 'social gifts' of cups of coffee for each other.

Currently nearly all at-home consumption is Robustabased instant products, with half of all imports coming from Vietnam. The projected development of the country has seen major coffee companies such as Nestlé and Starbucks working with the Chinese Arabica farmers in Yunnan province in anticipation of offering local products to this growing market.

Brazil is the outstanding example of a producer country that has developed a consumer market. Some 95 per cent of adults consume coffee, and the country is close to overtaking the United States as the largest national market. All coffee sold in Brazil must have been grown in the country.

Annual average consumption rates have doubled between the 1990s and the 2010s. A driver for this is the rise of a new middle class that makes up nearly half the population. They have embraced the arrival of the new-style coffee shops serving cappuccinos, with out-of-home consumption rising by 170 per cent between 2003 and 2009. The introduction of independently certified quality labels for roast and ground coffees has transformed domestic consumption, raising both volume and quality. [1]

Roasters, retailers and coffee shop chains using local beans have sprung up in Indonesia, the Philippines and Vietnam, currently the fastest- growing consumer markets in Asia. Vietnam's Trung Nguyên Corporation operates five processing plants producing soluble products that it exports to over sixty countries. It owns or supplies over a thousand coffee shops in Vietnam. Most stores offer beverages brewed with Robustabased powders and condensed milk, retaining their appeal to the local customer base.

The original tiger economies have moved on in their tastes. In Singapore, a well-established third wave culture reflects the city's global status. South Korea is currently the fastest-growing market for commercial espresso machines: Seoul supposedly has more coffee shops per capita than Seattle. Customers use them to escape their small apartments in the evenings, resulting in average visit lengths of over an hour, and opening times from 10 a.m. to 11 p.m.

A New Era for Coffee?

The specialty movement developed as a countervailing tendency against coffee's commoditization in the developed world. Its success can be judged from the turnaround in

---

[1] International Coffee Organization, *A Step-by-step Guide to Promote Coffee Consumption in Producing Countries* (London, 2004), pp. 154–207; 'Brazil', www.thecoffeeguide.org, March 2011.

U.S. per capita consumption, which returned to nearly 4.5 kilograms (10 lbs) in 2014.

Transnational corporations continue to dominate the industry, but there have been some significant shifts in their composition and character. Nestlé remains the largest roaster in the world, but its most dynamic element is Nespresso, which is positioned as a high-end specialty product. jab, the Luxembourg-based private equity company whose portfolio of coffee brands includes JDE (Jacobs Douwe Egberts), has also invested in specialty, acquiring Peets and the 'third wave' chains Intelligentsia and Stumptown, as well as Keurig.

Coffee consumption is now much more evenly spread around the globe, breaking down the binary division between producer and consumer continents. Europe remains the largest continent with roughly a third of the market, but Asia, North America and Latin America each now command around a fifth. Given that per capita consumption rates in Asia are around one-tenth of those in North America, the potential for further development can be appreciated.

The impact of the specialty revolution is laying the foundations for a new era of coffee. Between 2014 and 2016, the world's annual consumption of coffee exceeded the quantity of beans produced. In the medium to long term the likelihood is that this trend will continue as demand for coffee, fuelled by the growth of new markets, increases, while output falls because of economic and environmental factors.

Economic development not only spurs coffee consumption, but impacts production. Coffee growers around the world are getting older as their sons and daughters migrate into the cities in search of better opportunities. This may lead to a decline in output, but could also address the issue of smallholdings becoming unsustainable when they are divided up between family members.

Regional Distribution of World Coffee Consumption, 2012–16[1]

Region	Proportion of World Total (%)	Compound Annual Growth Rate (%)
Europe	33.3	1.2
Asia and Oceania	20.9	4.5
North America	18.4	2.5
South America	16.6	0.4
Africa	7.0	0.9
Central America	3.5	0.7
World	100	1.9

The biggest threat to coffee growing is climate change. It is estimated that there will be a 50 per cent reduction in the global area suitable for coffee production by 2050. [2] In addition to exceeding temperature extremes, increased climate volatility can affect yields through changes in rainfall patterns, disease and pest populations. While climate change may lead to new regions emerging—coffee is already being

[1] Data courtesy of ICO.

[2] World Coffee Research, *The Future of Coffee: Annual Report 2016*, www.worldcoffeeresearch.org.

cultivated in southern California—the impacts on the traditional growing areas and the farmers they support will be profound. Scientific programmes to breed more climate-resistant varieties that retain good flavour profiles may help to mitigate these effects, but many farmers will still need to relocate or switch into other crops.

Such changes in the fundamentals underpinning the coffee market could potentially strengthen prices, particularly to those producers able to access the specialty market. They have already changed the geopolitical structures of the industry. For consumers, however, the benefits of the specialty revolution are that they can enjoy a greater quality and variety of coffee than has ever before been available. Happy brewing!

Recipes

It is easy to be intimidated into thinking that preparing a great cup of coffee requires lots of high-tech equipment, knowledge and a well-stuffed wallet. Here are some tips to 'up your coffee game' at home without too much time or money.

Freshness is key. The biggest change you can make is to start buying whole beans and grinding them just before use. Ideally use an electric burr grinder that can be calibrated to grind equally sized particles. Even an inexpensive single-blade 'knife' grinder will make a huge difference.

Purchase small amounts of beans and use them before they go stale. Check the bag for the coffee's roasting date (not the 'best before' date)—it should be no more than three weeks old. Store at room temperature in a dry, dark place—*not* the fridge!

You can buy beans in a supermarket, but try a specialty supplier by doing a web search on 'artisan coffee [your town]', or sign up for an online subscription service. Purchase 'single origin beans' from an identified country, preferably a named region such as Ethiopian Yirgacheffe.

If possible, use digital scales to weigh the coffee and water. Coffee scoops normally hold around 10 g ground coffee. Water temperature should be between 90 and 95°c; a simple tip is to boil a kettle, then wait thirty seconds before using the water.

Suggested coffee and water brewing ratios
(approximate—dependent on taste)

Brewing Method	Serves	Ground Coffee (g)	Water (ml)	Brew time (mins)
French press	1	20	300	4
V60	1	18	250	3.5
Chemex	2–3	30	500	4
AeroPress (short)	1	15	150	0.5
AeroPress (long)	1	18	250	3.5
Espresso (shot)	1–2	7–9	25	25 seconds

French Press or Cafetière

The easiest and most forgiving way of brewing coffee. Grind the coffee coarse (think breadcrumbs), put it in the pot, add the heated water, cover, leave for four minutes (use a timer), then press the plunger down. Pour yourself a cup of flavourful coffee—perfect for breakfast.

Filter (V60, Chemex)

For a clean-tasting afternoon coffee, try a filter brewer such as the V60 or Chemex. Dampen the filter, add medium-ground coffee (think salt), shake level, pour a small amount of heated water to wet the coffee and wait for it to swell up or 'bloom'. Then gently pour over the remaining water at intervals and allow to drip through. Aim for two and a half minutes for a V60, four minutes for the Chemex.

AeroPress

The AeroPress is a great, portable device. Wet filter paper, attach filter basket to tube, stand on a mug, add medium-fine ground coffee then heated water, stir for ten seconds, insert plunger, push water through the coffee at a steady rate. For a concentrated cup, close to Americano, use AeroPress's scoop (approx. 15 grams) and add water to level number two. For a body between French press and filter, insert plunger into top of empty tube, turn upside down, add coffee and water, stir, steep three minutes, place mug over the top, turn over, press through slowly. Remove filter basket and blast the spent puck of coffee grounds out with the plunger.

Moka (Stovetop Espresso)

The secret to the stovetop Moka pot is to fill the bottom chamber to just below the air valve, loosely pack fine-ground coffee into the basket and turn the heat *off* once air is heard spluttering through the coffee into the upper chamber.

Espresso

Don't expect to make good espresso without investing heavily in machine and grinder. The grind (think sand) is vital because it controls the flow of the coffee: pre-ground espresso blends work better in a Moka. The basic principle of milk foaming is to insert the steam wand tip just under the surface of the milk, develop a vortex and then 'stretch' the milk by slowly lowering the milk jug, keeping the wand tip just under the surface of the milk. Coffee geeks love playing with their machines—you may find it quicker to visit your local coffee shop.

The Coffee Shop Menu

The international coffee shop menu is dominated by espressobased beverages, in combination with milk that is either steamed or frothed (foamed). Milk foam varies from the highly airy macrofoam used in a 'dry' cappuccino to the velvety micro-foam with tiny bubbles used in the flat white. Serving sizes vary from country to country and chain to chain.

Americano	Hot water topped with an espresso shot
Babyccino	Frothed milk without coffee prepared for children
Caffè Latte	Espresso shot topped with steamed milk and a small head of foamed milk. Syrups are added to create flavours such as gingerbread, pumpkin or vanilla
Cappuccino	Espresso shot topped with equal portions of steamed and foamed milk. The domed head can be dusted with cocoa or cinnamon
Cold Brew	Long summer drink obtained by infusing cold water and coffee grounds in refrigerated conditions for between 16 and 24 hours

Cold Drip	Long drink obtained by filtering cold water through coffee at a very slow rate—typically eight hours
Cortado	Spanish-style espresso shot (longer and weaker than Italian), topped with an equal amount of steamed milk
Espresso	Concentrated shot of coffee, 25–30 ml brewed under approx. 9 bars of pressure. Many coffee shops now use a double espresso shot as standard
Flat White	The third wave's favourite milky coffee drink originated in Australasia. Microfoamed milk is poured into a coffee cup containing a double ristretto, and the 'flat' top is usually finished with latte art
Iced Coffee	Regular brewed coffee chilled and served with ice
Macchiato	Espresso topped with a dash of foamed milk
Mocha	There are multiple variations, but the base elements are espresso, chocolate and steamed milk. Often served with marshmallows or similar additions
Nitro Coffee.	Nitrogen-infused cold-brew coffee resulting in a creamy taste
Piccolo	Espresso shot topped with an equal proportion of micro-foamed milk
Ristretto	Short, strong espresso, approx. 15 ml, common in southern Italy

Coffee has long been used as an ingredient in food and cocktails. Here are a few historical recipes that demonstrate its versatility. All measures and instructions are as in the original.

Coffee Cake

From Mrs Beeton's *Book of Household Management* (1861)

½ lb butter

½ lb brown sugar

¼ lb golden syrup

½ lb currants

1 lb sultanas

1½ lb flour

1 oz. baking powder

2 eggs

½ oz. mix of nutmeg, cloves and cinnamon

coffee

a little milk

Sieve the baking powder and spices with the flour into a bowl; add the sugar and butter, rub well together, make a well in the centre, pour in the syrup, add about ¼ pint of strong, cold, coffee, break in the eggs, and beat well together; then mix in the other ingredients with a strong wooden spoon using a little milk if not moist enough, mix in the fruit last, and then bake in a long square cake-pan nicely prepared. Bake from 1 to 2 hours. Sufficient for a cake about 1¾ lb.

Coffee Ice Cream

From A. Escoffier, *A Guide to Modern Cookery* (1907)

Ice-cream preparation

Work ⅔ lb. of sugar and 10 egg yolks in a saucepan until the mixture reaches the ribbon-stage. Dilute it, little by little, with one quart of boiling milk, and stir over a moderate fire until the preparation veneers the withdrawn spoon. Avoid boiling, as it might decompose the custard. Strain the whole into a basin and stir it from time to time until it is quite cold.

To freeze an ice preparation . . . surround it with broken ice, mixed with sodium chloride (sea-salt or freezing salt) and saltpetre. The action of these two salts upon the ice causes a considerable drop in the temperature which speedily congeals any contiguous liquid . . . The freezer [i.e. the receptacle], in which the freezing takes place . . . should be of pure tin . . . pour into it the preparation to be frozen and then either keep it in motion by rocking the utensil to and fro, by grasping the handle on the cover . . . or by turning the handle if the utensil is on a central axle . . . the rotary movement of the utensil causes the preparation to splash continually against the sides of the freezer, where it rapidly congeals, and the congealed portions are removed by means of a special spatula, as quickly as they form, until the whole becomes a smooth and homogeneous mass.

Coffee Flavouring

Add 2 oz. of freshly grilled and crushed coffee seeds to the boiled milk, and let them infuse for 20 minutes. Or, with an equivalent amount of ground coffee and ½ pint of water, prepare a very strong infusion and add it to 1½ pints of boiled milk.

Tiramisu

Tiramisu is considered a classic Italian dessert, but it was created in a Treviso restaurant in the 1970s. Here is the original restaurant recipe, available at www.tiramesu.it, which uses *savoiardi* biscuits (ladyfingers or sponge fingers).

12 egg yolks

500 g white sugar

1 kg mascarpone cheese

60 *savoiardi*

coffee

cocoa

powder

Make coffee, set aside and let cool in a bowl. Whip egg yolks with sugar until stiff; fold mascarpone into the mixture to create a soft cream. Dip 30 *savoiardi* into the coffee, being careful not to soak them. Arrange them in a line, in the middle of a round dish. Spread half of the cream over the *savoiardi*. Repeat using remaining *savoiardi* and cream to create a second layer on top of the first. Sprinkle with cocoa powder and serve chilled.

Bruleau, aka Brûlot, aka Café Diabolique

From Martha McCulloch-Williams, *Dishes and Beverages of the Old South* (1913)

Put into the special bruleau bowl, which has its own brandy ladle, three ladlefuls of brandy, along with the yellow peel of half an orange, a dozen cloves, a stick of cinnamon, a few grains of allspice and six lumps of sugar. Let stand several hours to extract the essential oils. At serving time put in an extra ladleful of brandy for every person to be served, and two lumps of domino sugar. Pour alcohol in the tray underneath the bowl, light it, and stir the brandy back and forth until it also catches from the flame below. Let burn two or three minutes—if the lights be extinguished as they should be, the effect is beautifully spectral. After the three minutes pour in strong, hot, clear, black coffee, a small cupful for each person, keep stirring until the flame dies out, then serve literally blazing hot. This 'burnt water' known in more sophisticated regions as *Café Diabolique*, originated in New Orleans, and is the consummate flowering of Creole cookery.

Irish Coffee

When a transatlantic flying boat arrived into Foynes airport in Ireland in the 1940s, the airport barman added whiskey to his coffee to warm the passengers up. Asked if it was Brazilian coffee, he replied, 'No, Irish.' The recipe subsequently evolved in Ireland and the United States. Here is the version recommended by the Pan-American Coffee Bureau in its 1956 publication *Fun with Coffee*.

Into a warmed table wine glass, place 2 teaspoons of white sugar and fill glass about two-thirds with coffee. Mix. Add about 2 tablespoons of Irish whiskey and top with softly whipped cream. (To float the cream on top of the coffee, try pouring it over the back of a spoon—do not stir once in place.)

Espresso Martini

30 ml (1 shot) espresso

50 ml (2 fl. oz) vodka

10 ml (2 tsp) sugar syrup

Place all ingredients into a cocktail shaker filled with ice, and shake for at least 10 seconds. Pour into a chilled martini glass and garnish with three coffee beans placed close to each other.

Select Bibliography

Clarence Smith, William Gervase, and Steven Topik, eds, *The Global Coffee Economy in Africa, Asia, and Latin America, 1500–1989* (Cambridge, 2003)

Daviron, Benoit, and Stefano Ponte, *The Coffee Paradox* (London, 2005)

Ellis, Markman, *The Coffee House: A Cultural History* (London, 2004)

Folmer, Britta, ed., *The Craft and Science of Coffee* (Amsterdam, 2017)

Hattox, Ralph, *Coffee and Coffeehouses: The Origins of a Social Beverage in the Medieval Near East* (Seattle, WA, 1985)

Laborie, P. J., *The Coffee Planter of Saint Domingo* (London, 1798)

Maltoni, Enrico, and Mauro Carli, *Coffeemakers* (Rimini, 2013)

Palalcios, Marco, *Coffee in Colombia 1850–1970* (Cambridge, 1980)

Prendergast, Mark, *Uncommon Grounds* (New York, 2010)

Roseberry, William, Lowell Gudmondson and Mario Samper Kutschbach, eds, *Coffee, Society and Power in Latin America* (Baltimore, MD, 1995)

Talbot, John M., *Grounds for Agreement* (Lanham, MD, 2004)

Thurston, Robert, Jonathan Morris and Shawn Steiman, eds,
Coffee: A Comprehensive Guide to the Bean, the Beverage and the Industry (Lanham, MD, 2013)

Ukers, William H., *All About Coffee* (New York, 1935)

Vidal Luna, Francisco, and Herbert S. Klein, *The Economic and Social History of Brazil since 1889* (Cambridge, 2014)

Websites and Associations

Allegra World Coffee Portal
www.worldcoffeeportal.com

Comunicaffe International
www.comunicaffe.com

Global Coffee Report
www.gcrmag.com

International Coffee Organization
www.ico.org

National Coffee Association
www.ncausa.org

Perfect Daily Grind
www.perfectdailygrind.com

Specialty Coffee Association
www.sca.coffee

Tea and Coffee Trade Journal

www.teaandcoffee.net

World Coffee Research

www.worldcoffeeresearch.org

Acknowledgements

Writing this book has been like brewing an espresso: a huge amount of material has had to be concentrated down to produce an easily consumed final shot.

I'd like to thank Michael Leaman for commissioning this volume and giving me the time to prepare it properly, and the staff at Reaktion for all their assistance. Allegra Strategies, Comunicaffè, the International Coffee Organisation, the MUMAC Academy, the Specialty Coffee Association and World Coffee Research have all generously shared information with me, while I continue to be astounded by the generosity of the many members of the coffee industry from whom I've learned along the way. I am especially grateful to Ago Luggeri, Anna Hammerin, Anya Marco, Arthur Ernesto Darboven, Barbara Derboven, Barry Kither, Britta Folmer, Clive Maton, Colin Smith, Darcio De Camillis, Enzo Frangiamore, Ender Turan, Enrico Maltoni, Kenneth McAlpine, Kent Bakke, Lindsay Eynon, Luigi Morello, Maurizio Giuli, Robert Thurston, Shawn Steiman and Yasmin Silverman for their advice on specific aspects of this text.

This book would never have seen the light of day without the support of my wife, Elizabeth. I am blessed to share my coffee with her.

Photo Acknowledgements

The author and publishers wish to express their thanks to the below sources of illustrative material and/or permission to reproduce it. Some locations of artworks are also given below, in the interests of brevity:

Bettmann/Contributor/Getty Images: p. 119; © Château des ducs de Bretagne–Musée d'histoire de Nantes: p. 84; DeAgostini/G. DAGLI ORTI/ Contributor: p. 103; Elizabeth Dalziel: p. 41; Peter Harris/ SteamPunkCoffeeMachine: p. 9; International Coffee Organization: pp. 19, 33, 109, 112, 123, 125, 131, 144; Keystone-France/Gamma-Keystone via Getty Images: p. 107; Lebrecht Music and Arts Photo Library/Alamy Stock Photo: p. 76; Library of Congress, Washington, DC: pp. 44, 90; Stuart McCook: pp. 96, 99; Jonathan Morris: pp. 23, 26, 35, 37, 51, 137, 160, 162; © Nespresso: p. 165; courtesy Nestlé Historical Archives, Vevey, Switzerland: p. 127; Jake Olson for World Coffee Events: p. 161; © Paulig Coffee: p. 133; John Phillips/The LIFE Picture Collection/ Getty Images: p. 105; Steve Raymer/CORBIS/Corbis

License.

Readers are free:

to share—to copy, distribute and transmit the work
to remix—to adapt this image alone

Under the following conditions:

attribution—You must attribute the work in the manner
specified by the author or licensor (but not in any way
that suggests that they endorse you or your use of the
work).

share alike—If you alter, transform, or build upon this
work, you may distribute the resulting work only under
the same or similar license to this one.

图书在版编目（CIP）数据

咖啡 /（英）乔纳森·莫里斯著；赵芳译 . —— 北京：北京联合出版公司，2023.10
（食物小传）
ISBN 978-7-5596-7155-4

Ⅰ．①咖… Ⅱ．①乔… ②赵… Ⅲ．①咖啡－文化史－世界－普及读物 Ⅳ．① TS971.23-49

中国国家版本馆 CIP 数据核字（2023）第 147556 号

咖　啡

作　　者：〔英国〕乔纳森·莫里斯
译　　者：赵　芳
出 品 人：赵红仕
责任编辑：夏应鹏
产品经理：马　婷
装帧设计：鹏飞艺术
封面插画：〔印度尼西亚〕亚尼·哈姆迪

北京联合出版公司出版
（北京市西城区德外大街 83 号楼 9 层　　100088）
北京天恒嘉业印刷有限公司印刷　　新华书店经销
字数 183 千字　889 毫米 ×1194 毫米　1/32　12.5 印张
2023 年 10 月第 1 版　　2023 年 10 月第 1 次印刷
ISBN 978-7-5596-7155-4
定价：69.00 元

版权所有 侵权必究
北京市版权局著作权合同登记　图字：01-2022-7179 号